Joseph Henry Wythe

The Physician's Dose and Symptom Book

Joseph Henry Wythe

The Physician's Dose and Symptom Book

ISBN/EAN: 9783337345617

Printed in Europe, USA, Canada, Australia, Japan

Cover: Foto ©berggeist007 / pixelio.de

More available books at **www.hansebooks.com**

THE
PHYSICIAN'S
DOSE AND SYMPTOM BOOK,

CONTAINING

THE DOSES AND USES

OF ALL

THE PRINCIPAL ARTICLES OF THE MATERIA MEDICA AND OFFICINAL PREPARATIONS;

ALSO,

Table of Weights and Measures.	Index of Diseases and Treatment.
Rules to Proportion the Doses of Medicines.	Pharmaceutical Preparations.
Common Abbreviations used in Writing Prescriptions.	Table of Symptomatology.
Table of Poisons and Antidotes.	Outlines of General Pathology and Therapeutics.

BY

JOSEPH H. WYTHES, A. M., M. D.,

LATE SURGEON U. S. VOL'S; AUTHOR OF "THE MICROSCOPIST," ETC., ETC.

ELEVENTH EDITION, REVISED.

PHILADELPHIA:
LINDSAY & BLAKISTON.
1874.

Entered according to Act of Congress, in the year 1874, by
LINDSAY & BLAKISTON,
in the Office of the Librarian of Congress at Washington.

PRINTED BY SHERMAN & CO.

PREFACE

TO THE ELEVENTH EDITION.

THE number of copies sold of previous editions of this vade mecum prove its utility. The present edition has been carefully revised and compared with the new U. S. Dispensatory, etc., so as to embody the recent additions to the Materia Medica, as well as every article likely to be useful.

The endeavor has been to make it a book full of suggestions, as well as to save the trouble of reference to more elaborate works.

DOSE AND SYMPTOM BOOK.

I.

TABLE OF WEIGHTS AND MEASURES.

APOTHECARIES' WEIGHT.

Pound.	Ounces.	Drachms.	Scruples.	Grains.
℔ 1 =	12 =	96 =	288 =	5760
	℥ 1 =	8 =	24 =	480
		ʒ 1 =	3 =	60
			℈ 1 =	20

APOTHECARIES' OR WINE MEASURE.

Gallon.	Pints.	Fluidounces.	Fluidrachms.	Minims.
Cong. 1 =	8 =	128 =	1024 =	61440
	O 1 =	16 =	128 =	7680
		f℥ 1 =	8 =	480
			fʒ 1 =	♏ 60

The drop of water is about equal to a minim, 60 drops being equal to a fluidrachm. It is important, how-

TABLE OF WEIGHTS AND MEASURES.

ever, to bear in mind that this is by no means the case with other fluids; for example, a minim of the tinctures being about equal to two drops, or 120 drops to the fluidrachm.

DOMESTIC MEASURES.

A teacup is considered equal to 4 fluidounces.
A wineglass " " 2 "
A tablespoon " " ½ "
A teaspoon " " 1 fluidrachm.

II.

RULES TO PROPORTION THE DOSES OF MEDICINE.

In prescribing, the following circumstances should always be kept in view: — Age, Sex, Temperament, Habit, Climate, the Condition of the Stomach, and Idiosyncrasy.

Age. — For an adult, suppose the dose to be ONE, or 1 drachm. Under 1 year will require only 1-12th, or 5 grains. Under 2 years will require only 1-8th, or $7\frac{1}{2}$ grains. Under 3 years will require only 1-6th, or 10 grains. Under 4 years will require only 1-4th, or 15 grains. Under 7 years will require only 1-3d, or 1 scruple. Under 14 years will require only $\frac{1}{2}$, or $\frac{1}{2}$ drachm. Under 20 years will require only 2-3ds, or 2 scruples. Above 21 years the full dose, 1 drachm. Above 65, the inverse gradation of the above.

Opiates affect children more powerfully than adults, but children bear larger doses of calomel than adults.

Sex. — Women require smaller doses than men; they are more rapidly affected by purgatives than men, and the condition of the uterine system must never be overlooked.

Temperament. — Stimulants and purgatives more readily affect the sanguine than the phlegmatic, and consequently the former require smaller doses.

Habits. — The knowledge of habits is essential; for persons in the habitual use of stimulants and narcotics, require larger doses to affect them when laboring under disease, while those who have habituated themselves to the use of saline purgatives, are more easily affected by these remedies. Persons, however, who have habituated themselves to the use of opium, do not require larger doses than usual of other narcotics.

Climate. — Medicines act differently on the same individual in summer and in winter, and in different climates. Narcotics act more powerfully in hot than in cold climates, hence smaller doses are required in the former; but the reverse is the case with regard to calomel.

Condition of the Stomach, and Idiosyncrasy. — The least active remedies operate very violently on some individuals, owing to a peculiarity of stomach, or rather disposition of body, unconnected with temperament. This state can be discovered only by accident or time; but when it is known, it should always be attended to.

In prescribing, the practitioner should always so regulate the intervals between the doses, that the next dose may be taken before the effect produced by the first is altogether effaced; for, by not attending to this circumstance, the cure is always commencing, but never proceeding. It should, however, also be kept in mind, that some medicines, such as the mercurial salts, arsenic, etc., are apt to accumulate in the system; and danger may thence arise, if the doses too rapidly succeed one another. The action also of some remedies, elaterium and digitalis, for example, continue long after the remedy is left off; and therefore much caution is requisite in avoiding too powerful an effect, by a repetition of them even in diminished doses. Aloes and castor oil acquire activity by continued use, so that the dose requires to be diminished.

The following simple rule, given by Dr. Young, will be found very useful as a guide in prescribing for children.

For children under 12 years of age, the doses of most medicines must be diminished in the proportion of the age to the age increased by 12; thus at 2 years $\frac{1}{7}$ — viz. $\frac{2}{2+12} = \frac{1}{7}$: or, in other words, add the age to 12, and divide the sum by the age, and the product will be the proportion of the dose to that of an adult. At 21, the full dose may be given.

III.

COMMON ABBREVIATIONS USED IN WRITING PRESCRIPTIONS.

Abbreviation.	Latin.	English.
āā	Ana.	Of each.
Ad lib.	Ad libitum.	At pleasure.
Ad saturand.	Ad saturandum.	Until saturated.
Aq. ferv.	Aqua fervens.	Hot water.
Aq. tepid.	Aqua tepida.	Warm water.
Chart.	Chartula.	A small paper.
Coch. mag.	Cochlear magnum.	A tablespoonful.
Coch. parv.	Cochlear parvum.	A teaspoonful.
Colent.	Colentur.	Let them be strained.
Collyr.	Collyrium.	An eyewater.
Contus.	Contusus.	Bruised.
F. vel ft.	Fiat vel fiant.	Let them be made.
Fol.	Folium vel folia.	A leaf or leaves.
Garg.	Gargarisma.	A gargle.
Gtt.	Gutta vel guttæ.	A drop or drops.
Haust.	Haustus.	A draught.
Infus.	Infusis.	An infusion.
M.	Misce.	Mix.
Mass.	Massa.	A mass.
Mist.	Mistura.	A mixture.
Pil.	Pilula vel pilulæ.	A pill or pills.
Pulv.	Pulvis vel pulveres.	A powder or powders.
Q. s.	Quantum sufficit.	A sufficient quantity.
℞.	Recipe.	Take.
Rad.	Radix.	A root.
S.	Signa.	Write.
Ss.	Semis.	The half.
Tinct.	Tinctura.	A tincture.

IV.

TABLE OF POISONS AND ANTIDOTES.

In all cases of poisoning, the first step is to evacuate the stomach, which should be effected by one of those emetics which is most powerful and speedy in its operation, as sulphate of zinc, or sulphate of copper. When vomiting has already taken place, copious draughts of warm water or mucilaginous drinks should be given, to keep up the effect till the poisoning substance has been evacuated. If vomiting cannot be produced, the stomach-pump must be used.

Inflammation of the stomach, congestion of the brain, and other symptoms, are to be treated on general principles, viz., by bloodletting, cold applications, revulsives, cool mucilaginous drinks, milk, lime-water, etc. When prostration exists, stimulants should be resorted to, as in other cases.

The following is a list of the usual poisoning substances, with the appropriate remedies:

Poisons.	Antidotes.
Acids.	The Alkalies. Common soap (soft or hard) in solution is an efficient remedy, and has the advantage of being always at hand. It should be followed by copious draughts of tepid water or flaxseed tea. For nitric and oxalic acids, the carbonates of magnesia and lime (chalk and water) are the best antidotes. When sulphuric acid has been taken, the use of much water will be improper.
Alkalies and their salts.	The Vegetable acids. Common vinegar being always at hand, is most frequently used. The fixed oils, as castor, flaxseed, almond and olive oils, form soaps with the alkalies, and thus, also, destroy their caustic effect. They should be given in large quantities.
Earths. Baryta and its salts. Lime.	Epsom or Glauber's Salts, in solution, or diluted sulphuric acid. The fixed oils also have the same effect as with the alkalies proper when not in a compound state.
Iodine. Iodide of Potassium.	Starch, or wheat flour, in large quantities well mixed with water. For Iodide of Potassium, there being no antidote, vomiting must be pro-

Poisons.	Antidotes. — (Continued.)
	moted by copious draughts of warm water.
Antimony and its salts.	Astringent Infusions, as of galls, oak bark, Peruvian bark, or green tea, very strong.
Arsenic and its compounds.	Hydrated Peroxide of Iron, in tablespoonful doses every 5 or 10 minutes. See Ferri Oxidum Hydratum.
	Freshly precipitated Magnesia, Demulcents, etc.
Bismuth and its compounds.	Albumen. Copious draughts of milk, combined with sweet mucilaginous drinks.
Copper and its compounds.	Albumen, as milk or whites of eggs in solution, should be freely administered. Vinegar must be avoided.
Gold, salts of.	Sulphate of Iron, with a free use of mucilaginous drinks.
Iron, salts of.	Carbonate of Soda, with mucilaginous drinks.
Lead, salts of.	Sulphate of Magnesia (Epsom salts), or diluted sulphuric acid.
Mercury, salts of.	Albumen, as whites of eggs, milk, or wheat flour beaten up with water.
Silver, salts of.	Common salt (chloride of sodium), largely given.

Poisons.	Antidotes. — (Continued.)
Tin, salts of.	Albumen. Whites of eggs, milk, or flour.
Zinc, salts of.	Albumen, or carbonate of soda, with copious draughts of warm water, and especially milk.
Phosphorus.	Magnesia with water, and copious draughts of mucilaginous drinks.
Gases.	Ammonia cautiously inhaled is recommended for chlorine. Asphyxia produced by carbonic acid or carbonic oxide gases or sulphuretted hydrogen, must be treated by copious effusions of cold water, especially to the head, bloodletting, artificial respiration, stimulants carefully administered, etc.
Creasote.	Albumen, or whites of eggs, milk, or wheat flour.
Alcohol or spirituous liquors.	A powerful emetic should be given, followed by copious draughts of warm water. Congestion of the brain and other symptoms, to be treated on general principles.
Opium and other narcotics.	The chief reliance is to be placed on the most active emetics (as tartar emetic, sulphate of copper, or sulphate of zinc), and the stomach-pump. Emetics are preferable to the stomach-pump when the nar-

Poisons.	Antidotes.— (Continued.)
	cotic has been taken in substance. The patient should be kept in motion, and cold water dashed on the head and shoulders. Bloodletting may become necessary to relieve congestion. After other remedies fail, artificial respiration should be resorted to.
	Strong hot coffee, a teacupful alternately with vinegar and water may be useful.
	Electro-magnetism has also been efficacious.
Poisonous Serpents.	A cupping-glass over the wound, or a tight ligature above it; cauterization of wound. Warm diluent drinks and small doses of ammonia to promote perspiration. Whiskey has been given in large doses.
	Bibron's antidote is as follows: ℞. Potass. Iodid., gr. iv, Hydrarg. chlor. corros., gr. ij, Brominii ʒv. M. 10 drops in a tablespoonful of wine or brandy, repeated if necessary.

V.

INDEX OF DISEASES AND TREATMENT.
(After Tanner.)

Abscess.
1. *Of brain.* — Pain, vertigo, paralysis, etc. Treatment doubtful.
2. *Of external ear.* — Pain, swelling, etc. Fomentations, etc.
3. *Of tonsil.* — Chill, fever, swelling, pain. Open towards median line.
4. *Retro-pharyngeal.* — Fever, sore throat, swelling, etc. Puncture (Tracheotomy?).
5. *Empyema.* — Intercostal bulging, dull percussion, no murmur. Sinapism, Iod. Potas., Tapping, etc. Tonics.
6. *Hepatic.* — Chills, hectic, pain, prostration, etc. Tonics. Puncture if parietal adhesion.

General treatment of abscesses similar.

If strumous, chemical food (Phosphates of lime, iron, soda, and potass., in syrup) and Cod-liver oil.

Bladder Diseases.
1. *Irritability.* — Analyze urine chemically and mi-

croscopically, and trace disease to its source. Mucilaginous drinks, etc.
2. *Spasm.* — Pain, suppression, tenesmus, etc. Hot bath, Camphor or poppy fomentations to perineum, or lotion of Tinct. Aconite, etc.
3. *Paralysis.* — Retention, distension, pain, coma. Catheter, hip-bath, blisters, etc.
4. *Inflammation.* — Catheter, fomentations, baths, opium, aconite, mucilages, tonics, etc. If chronic, buchu, cubebs, opiate suppositories, etc.
5. *Tumors and calculi.* — Chloroform, hot baths, narcotics, astringents. Analyze urine. Examine surgically. Lithotomy.

Blood Diseases.

1. *Anæmia.* — Iron. Chemical food. Tonics.
2. *Chlorosis.* — Good food and air, sea bathing. Chalybeates. Aloetic aperients.
3. *Hyperæmia.* — Restricted diet, exercise, salines. Liq. potassa, Liq. arsenicalis, Tartar emetic, etc.
4. *Pyæmia* (absorption of pus). — Shivering, sweating, rapid pulse, diarrhœa, pleurisy, peritonitis, etc. Death from prostration. Vapor bath. Quinine. Sulphites. Tonics. Stimulants.
5. *Acholia* (toxæmia from arrest of bile). — Nervous excitement. Typhoid prostration. Coma. Hemorrhage. Sometimes jaundice. Purgatives. Croton oil. Hydrochlorate of ammonia. Nitro-hydrochloric acid.
6. *Glucohæmia* (*Diabetes mellitus*). — Symptoms in-

sidious, feverishness, debility, excessive flow of urine, ends in phthisis, or some low form of inflammation. Test urine for sugar. Diet free from starch, etc. Muriated Tinct. Iron, Citrate of Ammonia, and iron. Quinine, Pepsin, Aperients. Nitro-hydrochloric acid. Vapor bath. Flannel, etc.

7. *Uræmia* (Toxæmia from absorbed urea). — Disturbance of nerve-centres. Convulsions. Coma. Albuminuria. Suppression of urine. Vapor bath. Acid sponging. Saline aperients. Elaterium. Croton oil. Enema of castor oil and turpentine. Stimulants, etc.

8. *Thrombosis — Embolism* (fibrous coagula in blood-vessel). — Support the vital powers, and allay irritability.

9. *Scurvy.* — Lemon-juice. Fresh vegetables. Raw meat. Citric acid. Iron. Tannin, etc.

10. *Purpura.* — Tendency to effusion, petechia, debility, etc. Treat as for scurvy.

Bone Diseases.

1. *Caries.* — Constitutional disturbance. Deep-seated pain. Abscess. Fistulous openings. Probe leads to dead bone. Tonics. Iron. Cod-liver oil. Chemical food. Remove dead bone. Inject dilute Carbolic acid.

2. *Necrosis* (dead bone inclosed in case of new bone, necrosis of superficial layer is exfoliation). — Operation for removal of sequestrum.

3. *Periostitis.* — Pain and tenderness, worse at night. Nodes, etc. Calomel and opium. Iodides. Iron. Blisters. Belladonna and mercury liniment. Incisions.
4. *Rickets* (softening of bone in children). — Animal food. Chemical food. Hypophosphite of lime. Milk. Cod-liver oil. Salt bath, etc.
5. *Coccyodynia* (pain and tenderness of coccyx, often very chronic). — Remove uterine or ovarian disease. Nerve tonics. Leeches. Hip-baths. Hypodermic injection of morphia. Subcutaneous section of muscles, or removal of bone.

Brain Diseases.

1. *Apoplexy.* — Care for the predisposition by temperance, exercise, cold affusion, etc. Bleeding if pulse full, hard, and thrilling, but avoid it if tendency to syncope. Turpentine enema. Pediluvia, etc.
2. *Hydrocephalus.* — Tonic regimen. Cod-liver oil. Salt baths. Rhubarb and magnesia. Quinine. Iodide of Iron, etc. If fontanelle depressed as in spurious hydrocephalus, Chemical food, Raw meat, Beef tea, etc. Avoid purges.
3. *Concussion.* — Distinguish from compression by easy breathing, and natural pupils although insensible to light. In compression, there is snoring and dilated pupil; in poisoning by opium, etc., contracted pupil, livid countenance, etc. Stimulants till reaction. Cold to head. Croton oil. Enema. Rest, etc.

4. *Coup de soliel* (sun-stroke). — Faintness, vertigo, sickness, coma. Cold to head. Stimulants. Sinapisms to extremities. Stimulant enema. Ice to spine. Friction.
5. *Meningitis.* — Fever, pain, delirium, coma. Calomel and jalap. Croton oil, etc. Iod. Potass. Ice to head. Stimulants in collapse.
6. *Chronic Encephalitis* (softening, etc.). — Vertigo, pain, failure of intellect, delusion, constipation, etc. Hygienic measures. Cod-liver oil. Blisters behind ears, etc. Analyze urine, lest diabetes, etc., be mistaken for it.
7. *Alcoholism.* — A degrading vice for which the person is responsible. Cured by total abstinence. Tonics, as Quinine, Pepsine, etc. In delirium tremens, Ammonia, Ether, Valerian, Morphia, Hydrate of Chloral. Laxatives, etc.
8. *Insanity* (mania, monomania, dementia, or idiocy).— Rest of mind. Sleep. Attend to functional derangement.
9. *Headache.* — Treat according as it is organic, plethoric, bilious or nervous. Intermittent hemicrania requires Quinine, etc.
10. *Paralysis.* — May be general, hemiplegic, paraplegic, local, rheumatic, from mercury, from lead, shaking palsy, progressive muscular atrophy, or progressive locomotor ataxy. Each variety requires careful study. Most cases need Tonics.
11. *Epilepsy.* — In the fit, loosen dress, protect the tongue by a cork, etc. Cold affusion to head. In

the interval, Bromide of Potassium. Quinine. Iron. Zinc. Ice bags to spine, etc.

12. *Aphasia* (derangement of speech, often sudden, and with no real loss of intellect). — Recovery spontaneous if no hemiplegia, otherwise generally hopeless. Iod. Potass.

Female Diseases.

1. *Vulval Disease.* — Pruritus often needs general treatment and lotions of Acet. Plumbi, Borax, Morphia, and Glycerine, etc. Tumors require Surgery. Vulvitis needs cleanliness, hip-baths, Alum or lead lotions, morphia, etc. Infantile leucorrhœa is often strumous.

2. *Vaginal Disease.* — Occlusion, prolapsus, and tumors require Surgery. Vaginitis is acute or chronic. The first needs hot hip-baths, warm injections, pessaries of oxide of zinc and belladonna, or acetate of lead and opium, etc. The chronic (leucorrhœa), mineral acids, Quinine, Iron, astringent applications, as Tannin, etc., Nit. Silver in solution, or Carbolic acid in Glycerine (gr. 10 to ʒj).

3. *Uterine Disease.* — Cancer, displacements, tumors, etc., are surgical. Ulceration of the os needs various treatment. Extent, etc., ascertained with speculum. Nit. Silver in substance — sometimes passed in the neck. Pessaries of Iod. Lead, Belladonna, Tannin, etc. Tonics. — In Menorrhagia, ice in vagina or over pubes; injections of Tannin, etc.; Tampon; Acet. Lead, Opium, Alum, Nut-

meg, Cinnamon, Sulph. Acid, etc. — Metritis — Repose, low diet, hot hip-baths, Opium and belladonna pessaries, Sinapisms to epigastrium, Ice, etc.
4. *Ovarian Disease.* — Dropsical tumor may need ovariotomy; drugs worse than useless. Ovaritis treated same as Metritis. If abscess points in vagina open with trocar or bistoury. Chronic ovaritis is often tedious. Iod. Potass., Bromide of Ammonium, Iron, Quinine, etc.

Fevers.
1. *Continued Fevers.* — 1. Simple continued fever — indications are to moderate arterial excitement by saline laxatives, rest, and diet; to support the system when it flags; to obviate local inflammation and congestion, and to relieve urgent symptoms as they arise. 2. Typhus. Ventilation, disinfectants, etc. Tepid sponging. Cold to head. Beef tea. Stimulants, etc. Quinine, Iron, and Mineral acids during convalescence. 3. Typhoid — term often used improperly. Should be restricted to enteric fever. Treat similar to Typhus, but avoid aperients, and treat diarrhœa with astringents and opium. 4. Relapsing. Amends on 5th or 7th day and relapses about 14th. See simple fever.
2. *Intermittent Fevers.* — Quinine in the intermission. The chief types are Quotidian, Tertian, and Quartan.
3. *Remittent Fevers.* — Treatment same as Intermittent. In the severest form, or Yellow fever, treat

as Typhoid. Avoid Ammonia, and be cautious with Alcohol.

4. *Eruptive Fevers.* — 1. Roseolæ. Sometimes simulates measles, sometimes scarlatina. Citrate of magnesia. Warm baths, etc. 2. Measles. Rash in blotches. Catarrh, etc. Milk diet. Castor oil. Inunction, etc. Watch pulmonary complications. 3. Scarlet fever. Acetate and Carbonate of Ammonia, Inunction, Stimulants, Tonics, etc. Watch sequelæ, as ulceration of tonsils, abscess of ear, anasarca, Albuminuria, etc. 4. Chicken-pox. Eruption of pimples which on second day are transparent vesicles, forming scabs on fourth day. Pyrexia slight. 5. Small-pox. Incubation 12 days, then fever and great backache, and in 48 hours an eruption of small red pimples which in a week inflame and suppurate. — In simple cases the less drugging the better. In cases of depression Quinine, Stimulants, etc. Vaccination.

	Incubation.	*Eruption.*	*Fading.*
Measles	10–14 days.	4th day.	7th day of fever.
Scarlet fever	4–6 "	2d "	5th " "
Small-pox	12 "	2d "	Scabs on 9th or 10th fall off about 14th.

6. Erysipelas. Tincture of perchloride of iron. Quinine. Fomentations. Solution of Sulphate of iron. Nit. Silver, etc.

Heart Diseases.

Pericarditis known by friction sounds, etc.
Endocarditis by bellows murmur, etc. Valvular

disease by bellows or musical murmur—if synchronous with pulse and most audible at apex, mitral disease; if not synchronous and most audible over sternum and aorta, aortal disease.

Hypertrophy leads to increased impulse, dyspnœa, palpitation, etc. Dilatation produces debility. Functional derangement of heart from hysteria, anæmia, neuralgia, etc., very common. Requires Antispasmodics, Ether, Ammonia, etc.

Intestinal Diseases.

1. *Colic*— from indigestion and flatulence, Ether. Ammonia. Brandy. Vomiting. Purging. Castor oil and laudanum—from mineral poison, Sulphate Magnesia, Sulphuric Acid, etc.
2. *Diarrhœa.*— Castor oil and laudanum. Calomel, etc. Afterwards Astringents, etc.
3. *Obstruction.* — Castor oil. Croton oil. Enema. Fomentations. Manipulation.
4. *Duodenal Dyspepsia.* — Pain, faintness, etc., about three hours after eating. Mercury. Nitromuriatic acid, etc.
5. *Enteritis.* — Rest. Opium. Calomel. Fomentations, etc.
6. *Dysentery.* — Rest. Demulcent drinks. Farinaceous food. Poultices. Castor oil. Opiate enema and suppositories. Bismuth. Gallic acid, etc.
7. *Cholera.* — Isolation. Disinfectants. (Carbolic acid?) Sinapisms. Stimulants, etc.
8. *Hæmorrhoids.* — External, should be excised. In-

ternal removed by ligature. Attention to digestion, and laxatives relieve.

Kidney Diseases.
1. *Nephritis.* — Hot hip-baths. Fomentations. Mild aperients. Diaphoretics, etc.
2. *Chronic Nephritis* (Bright's disease, etc.). — Examine urine for Albumen. See *Uræmia*.
3. *Diabetes.* — See *Glucohæmia*.
4. *Chylous urine.* — Gallic acid. Iron, etc.
5. *Hæmaturia.* — Examine cause. Tinct. Iron. Gallic acid. Ice to loins, perineum, etc.

Laryngeal and Tracheal Diseases.
1. *Aphonia.* — Functional: Quinine, Iron,. Nux Vomica, etc. Organic: Nit. Silver. Spray of astringent fluids, etc.
2. *Laryngitis.* — Acute: Antiphlogistic treatment. Warm moist air. Inhalation of stramonium, belladonna, etc. Tracheotomy. Chronic: Nit. Silver. Inhalation of medicated spray. Alteratives. Tonics.
3. *Laryngismus stridulus.* — In paroxysm, warm bath with cold affusion to head and face. Chloroform, etc.
4. *Dysphonia clericorum* (Follicular disease of pharyngo-laryngeal membrane — often nervous). — Quinine and Iron. Cold shower-baths. Iod. Potass. Iodohydrargyrate of Potass. Inhalation of atomized alterative or astringent fluids. Spong-

ing the larynx with Nit. Silver (40–60 grs. of crystals to ℥j Water). Excise tonsils.
5. *Diphtheria* (Exudation of false membrane with low fever).— Inhalation of acid vapor. Solution of hydrochloric acid, chlorinated water, nitrate of silver, etc. Tinct. Iron and Quinine. Chlorate Potass. Ice, etc. Sometimes Tracheotomy.
6. *Croup.* — Spasmodic: Warm bath, Emetics. Membranous: Emetics, Calomel. Belladonna to throat outside. Inhalations of warm vinegar, etc. Tracheotomy.

Liver Diseases.
1. *Hepatic congestion.* — Passive: Sulphate and Citrate Magnesia. Senna. Taraxacum, etc. Active: Podophyllum. Nitro-hydrochloric acid. Aloes, Senna, and Jalap, etc.
2. *Hepatitis.* — Rest in bed. Fomentations. Restricted diet. Sulphate of Soda and Taraxacum. Opium, etc. When chronic: Nitro-hydrochloric acid. Hydrochlorate of Ammonia, etc.
3. *Biliary calculi.* — Hot baths. Fomentations. Morphia. Chloroform. Castor oil, etc.
4. *Jaundice.* — From suppression: Sulphate of Soda and Taraxacum. Podophyllum. Hydrochlorate of Ammonia, etc. From obstruction: Podophyllum. Aloes. Croton oil. Sulphate of Magnesia. Fel Bovinum, etc.

Lung Diseases.
1. *Catarrh.* — Warm bath. Dover's powder. Purge.

INDEX OF DISEASES AND TREATMENT.

2. *Influenza.* — Rest. Diet of slops. Inhalation of vapor, Iodine or Belladonna spray. Sulphate Magnesia and Senna. Diaphoretic mixture. Quinine, etc., after.
3. *Bronchitis.* — Acute: see above. — Chronic, to be treated according to its nature. If secretion excessive: Alum, Squills, Ammonia, etc. Inhalation of atomized fluids.
4. *Hooping-cough.* — Ipecac, if much mucus. Senega. Nitric acid. Tinct. Aconite (1–2 minims). Belladonna liniment to spine. Alum and ginger, etc.
5. *Asthma.* — During paroxysm, a stimulant emetic or enema. Ammonia and Ether. Stimulants. Tobacco. Datura Tatula cigars. Nitre-paper fumes, etc. In interval, tonics, shower-bath, etc. Inhalation of spray.
6. *Emphysema.* — Invigorating diet, warm clothing. Carb. Ammonia. Ether. Quinine. Iron. Stramonium smoking. Warm climate, etc.
7. *Pleurisy.* — Rest. Fomentations. Cupping. Aperients. Opium. Aconite. Cream of Tartar. Quinine, etc.
8. *Pleurodynia* (neuralgic pain in side). — Belladonna and opium liniment. Sinapisms, etc.
9. *Pneumonia.* — Acute: Rest. Moist air. Acetate of Ammonia. Opium. Tartarized Antimony. Veratrum viride. Fomentations, etc. Chronic: Iod. Potass. Iod. Iron. Hydrochlorate Ammonia. Cod-liver oil, etc.

10. *Pulmonary gangrene.* — Ammonia. Quinine. Iron, etc.
11. *Phthisis.* — Nutritious animal food. Stimulants. Cod-liver oil. Iron. Quinine. Inhalations, etc.
12. *Hæmoptysis.* — Mineral acids. Opium. Inhalation of perchloride of Iron. Alum, etc.

Stomach Diseases.

1. *Dyspepsia.* — Invigorating means. Rest. Early hours. Sea-bathing. Disuse of tobacco and alcohol. Avoid indigestible food. Pepsin. Iron. Oxgall. Nitro-hydrochloric acid. Bismuth. Quinine, etc. Bismuth, Bicarb. Soda, etc., for gastralgia. Cod-liver oil, etc., for bulimia.
2. *Gastritis.* — Acute: Ice. Enema. Morphia. Fomentations. Great care in diet. Chronic: Low diet. Ice. Bismuth, etc.
3. *Gastric catarrh* ("bilious attack" — "sick headache"). — Seidlitz powders. Rhubarb. Ipecac. Mercury and chalk. Sulphite of Soda. Bismuth, etc.
4. *Hæmatemesis.* — Rest. Ice. Alum. Tinct. Iron. Quinine and Iron, etc. Enema of beef tea and brandy.

Venereal Diseases.

1. *Balanitis.* — Cleanliness. Astringent lotions, etc.
2. *Gonorrhœa.* — Acute: Rest. Active purgatives. Mild astringent injections. Camphor (5 grs.) and belladonna ($\frac{1}{2}$ gr.) pill at bedtime for chordee.

Oil of yellow sandalwood, etc. Chronic (gleet): Temperance. Infusion of uva ursi. Injections of infusion of Hydrastis Canadensis, Alum, Tannin, etc. Lallemand's porte-caustic. Tonics, etc.

3. *Syphilis.* — 1. Primary. For indurated chancre, a mercurial course. Soft chancres, caustic of acid nitrate of mercury, Monsel's salts. Astringent lotions. Iron tonics and nourishing food. Phagedenic chancre: soothing lotions and poultices. Iron. Iod. Potass. and Sarsaparilla. Sloughing chancre: fomentations and poultices. Opium. Nourishment. Stimulants. 2. Constitutional: light nutritious diet. Warm baths. Blue pill. Calomel. Iodide of Mercury. Iodide of Potass., etc.

VI.

ALPHABETICAL LIST OF MEDICINES WITH THEIR USES, DOSES, ETC.

ABSINTHE. A compound French *liqueur*, consisting of alcohol, oil of wormwood, anise, etc. Effects more deleterious than alcohol, tending to epileptiform convulsions.

ABSINTHIUM. Wormwood. (*Artemisia absinthium.*)
Tonic, antispasmodic, anthelmintic, discutient, antiseptic.

Use. In intermittents, dyspepsia, gout, hypochondriasis, dropsy, and epilepsy not depending on organic changes. Clysters of the decoction are useful in ascarides.

Dose. In substance, ℈j to ℈ij. Infusion (℥vj to water Oj) f℥vj to f℥xij, three or four times a day.

Incompatible. Sulphates of iron and of zinc; acetate and diacetate of lead, nitrate of silver.

ACACIA. Gum Arabic. (*A. vera.*)
Démulcent, nutritious.

Use. In catarrh, pertussis, ardor urinæ, etc. Mucilage of Gum Arabic is often employed as a vehicle for other substances. To render them miscible, oils require three-fourths of their own weight, balsams and spermaceti equal parts, resin two parts, and musk five times its weight.

Dose. In substance, ℨss. to ℨij. In decoctions, ad libitum.

Incomp. Goulard's extract, alcohol, sulphuric ether, tincture of muriate of iron.

ACER PENNSYLVANICUM. Striped Maple.

Use. A decoction of bark in cutaneous affections, and of the leaves and twigs to relieve vomiting.

ACETIC ETHER. Stimulant and antispasmodic.

Externally, for rheumatic pain.

Dose. 15 to 30 drops.

ACETUM. Vinegar.

Refrigerant, diaphoretic, antiseptic, astringent; externally, stimulant, and discutient.

Use. In febrile complaints and scorbutus; it has been supposed to counteract the effects of opium and other narcotics, after the stomach has been completely cleared; but this is a mistake, and it should never be employed in such cases; steam of it inhaled in putrid sore throats and in scurvy; as a lotion in bruises, sprains, burns, and chronic ophthalmia. Antilithic, where the triple phosphates abound in the urine; diluted with water, it forms the best means of cleansing the eye of small particles of lime.

Vinegar is considered to be better adapted to extract the virtues of some vegetables than alcohol, and the list of such preparations might be enlarged with benefit.

Vinegar whey is made by stirring a small wine-glassful of vinegar and a dessert-spoonful of sugar in a pint of milk; boil 15 minutes and strain. Used as a drink in fevers.

Dose. fℨj to fℨiv. In clysters, fℨj to fℨij. Lotion. ℞. Aceti fℨj, Spiritus Ten. fℨiv, Aquæ fℨviij.

ACETUM AROMATICUM. Aromatic Vinegar. See *Acidum Aceticum Arom.*

ACETUM CANTHARIDIS. Vinegar of Cantharides. Cantharid. ℨij in Acid Acet. Oj. Macerate 8 days. A prompt vesicant.

ACETUM COLCHICI. Vinegar of Meadow Saffron. Colchici cormi recent. concisi ℨj. Aceti dist. fℨxvj. Spir. ten. fℨj.

Use. In ascites, hydrothorax, and gout.

Incomp. Alkalies, earths, alkaline and earthy carbonates, sulphuric acid.

Dose. fℨss. to fℨj, in any bland fluid.

ACETUM DESTILLATUM. Distilled Vinegar. (Distil one gallon of vinegar on a sand-bath, in a glass retort and receiver. Reserve the first seven pints for use.)

Refrigerant, slightly astringent.

Use. The same as that of vinegar; chiefly for pharmaceutical purposes. It is used in the form of vapor for purposes of fumigation, but it has no efficacy in destroying contagious or infectious matter. It is also a good addition in refrigerating lotions containing acetate of lead.

Dose. fℨj to fℨiv.

ACETUM LOBELIÆ. Vinegar of Lobelia. Lobelia in powder ℨiv, Dilute Acetic Acid Oij. Made by percolation or maceration.

Dose. 30–60 drops, as antispasmodic. fℨss. as emetic.

ACETUM OPII. Vinegar of Opium. Black drop.

℞. Coarsely powdered opium ℨv, Nutmeg ℨi, Sugar ℨviii, Dilute Acetic Acid q. s. Macerate with 1 pint of the dilute acid 24 hours, percolate, and add acid to make two pints.

Narcotic.

Use. A substitute for tincture of opium; it is less likely to affect the brain than the tincture.

Dose. ♏vij to ♏x.

ACETUM SANGUINARIÆ. Vinegar of Bloodroot. Sanguinar. pulv. ℨiv, Acid. acet. dil. Oij.

Dose. ♏xv–♏xxx as alterative expectorant.

ACETUM SCILLÆ. Vinegar of Squill. Powdered squill ʒiv, Dilute Acetic Acid q. s. Percolate 2 pints, or macerate for seven days, express and filter.

Diuretic, expectorant, emetic, in large doses purgative.

Use. In dropsies, asthma, and chronic catarrh.

Dose. ♏xv–ʒi, in cinnamon water or mint water.

ACHILLEA MILLEFOLIUM. Milfoil. Yarrow.
Mild aromatic, astringent.

Dose. Used in infusion, or 20 drops of volatile oil.

ACIDUM ACETICUM. Acetic Acid.
Stimulant, rubefacient, escharotic.

Use. Applied to the nostrils in syncope, asphyxia, and headache; destroys corns and warts.

Incomp. Alkalies, earths, alkaline and earthy carbonates.

ACIDUM ACETICUM AROMATICUM. Aromatic Vinegar. (Rosmarini sic. folior. Origani. sing. ʒj. Lavandulæ sic. ʒiv. Caryophyllorum cont. ʒss. Acidi Acetici Ojss. Macerate seven days, and filter the expressed liquor through paper.)

Odor, pungent and aromatic.

Use. As a grateful perfume in sick-rooms.

ACIDUM ACETICUM CAMPHORATUM. Acid. Acet. fʒx. Camph. ʒj. Alcohol fʒj.

ACIDUM ACETICUM DILUTUM. Diluted Acetic Acid. (1 part to 7 parts of water.)

100 grs. are saturated by 7.6 grains of crystallized Bicarb. of Potassa.

ACIDUM ARSENIOSUM. Arsenious Acid. (See p. 12.)

Use. To prepare the arsenical solution.
(See *Liquor Arsenicalis.*)

ACIDUM BENZOICUM. Benzoic Acid.
Stimulant; as an expectorant, doubtful; errhine.
Use. In chronic catarrh, but of very little efficacy.
Dose. Grs. x to ʒss.

ACIDUM CARBAZOTICUM. Picric Acid.
Formed by the action of Nitric acid on animal and vegetable substances.
Use. Tonic, astringent, antiperiodic.

ACIDUM CARBOLICUM. Carbolic Acid. Phenic Acid. Phenol.
A hydrated oxide of phenyl, produced in the manufacture of coal-gas. It has the taste and smell of Creasote.
A most complete disinfectant.
Use. In Surgery a mixture of Carbolic Acid and Oil, etc., has been found of great use in arresting the formation of pus, etc. A solution of 2–10 grs. to ℨj Water has also been of benefit as a lotion, gargle, etc., in putrid and eruptive diseases. It is one of the most valuable contributions of modern science.
Glycerin is an excellent menstruum for either internal or external use.
Dose. 1–2 drops of liquefied acid; or 1 gr. to grs. iv to ℨj water by means of the atomizer; or 1 part in 4 to 8 for external use.

ACIDUM CHROMICUM. Chromic Acid.
Crystallized from a mixture of Bichromate of Potass. and Sulphuric acid.
Use. As an escharotic.

ACIDUM CITRICUM. Citric Acid.
Refrigerant, antiseptic.
Use. In febrile and inflammatory complaints, and scorbutus; and dissolved in water, instead of recent lemon-juice, for the effervescing draught. (Proportion, ʒixss. to water Oj.)

Dose. Grs. x to ʒss., dissolved in water or any bland fluid.

Incomp. Sulphuric acid, nitric acid, acetate of lead, nitrate and acetate of mercury, alkalies, alkaline sulphurets.

ACIDUM GALLICUM. Gallic Acid.
Astringent.
Use. In uterine and vesical hemorrhages.
Dose. Grs. v to xv.

ACIDUM HYDRIODICUM DILUTUM. Diluted Hydriodic Acid. An aqueous solution of the acid gas.
Use. Same as Iodine.
Dose. fʒ diluted with water, etc.

ACIDUM HYDROCHLORICUM. (Acidum Muriaticum, U. S.)
Tonic, antiseptic, diuretic.
Use. In typhus; cutaneous eruptions; in gargles in inflammatory and putrid sore throat; in injections in gonorrhœa.
Incomp. Alkalies, earths, and their carbonates; metallic oxides, sulphuret of potassium, tartrate of potassa, tartar emetic, and most metallic salts.

ACIDUM HYDROCHLORICUM DILUTUM. Diluted Hydrochloric Acid. (Acidi Hydrochlorici f℥iv, Aquæ destillatæ f℥xij.) fʒj should saturate grs. 32 of crystallized carbonate of soda.
Dose. ♏x to ♏xx, properly diluted; in gargles, fʒss. to fʒij in fʒvj of fluid; injection, ♏viij, to water f℥iv.

ACIDUM HYDROCYANICUM DILUTUM. Diluted Hydrocyanic Acid. Cyano Hydric Acid, Prussic Acid. (Potassi Ferrocyanidi ℥ij. Acidi Sulph. ℥jss., Aq. Dist. Oiss.)

(Prussic Acid may be prepared for immediate use in the following manner: Take of Cyanide of Silver grs. 50½, Muriatic acid grs. 41. Distilled Water ℥j. Mix the muriatic acid with the distilled water, add the cyanide of silver, and shake the whole in a well-stopped vial. When the insoluble matter has subsided, pour off the clear liquor and keep it for use.)

Sedative, antispasmodic.

Use. In spasmodic coughs, asthma, hooping-cough, nervous affections, hiccough, palpitation of the heart, and in allaying the irritability of the stomach in dyspepsia. Prussic acid may be employed with great benefit in cases of chronic neuralgic affections of the stomach. It should be given in increased doses, till some physiological effects are produced; then continued in rather a diminished quantity. As a local application, properly diluted, it is useful in abating the itching in Impetigo and pruriginous affections.

Dose. ♏ij gradually increased to ♏v in a glassful of water, almond emulsion, or infusion of cinchona. When an overdose has been taken, the effects are best counteracted by ammonia, chlorine, brandy, and the cold affusion.

Incomp. Metallic oxides, chlorine.

ACIDUM LACTICUM. Lactic Acid.

Use. In dyspepsia, and phosphatic deposits.

Dose. ʒi to ʒiij in sweetened solution, or in connection with pepsin.

ACIDUM NITRICUM. Nitric Acid.

Tonic, antiseptic, antisyphilitic, escharotic.

Use. The strong acid is seldom used for any other than pharmaceutical purposes; in the form of vapor is extracted from nitre ʒiv, and sulphuric acid ʒiv, in a saucer, placed on a pipkin of hot sand, for the purpose of fumigation.

Incomp. Spirit of lavender and the strong tinctures, in any large quantity; and the essential oils; metallic oxides.

ACIDUM NITRICUM DILUTUM. Diluted Nitric Acid.

Use. As a drink, diluted largely, in fevers of the typhoid kind; in chronic affections of the liver, attended with a redundant and hasty formation of bile, and in dyspepsia. It is also very useful in the cure of old ulcerated legs.

Dose. ♏x to ♏xx in f℥iij of water, twice or thrice a day.

ACIDUM NITRO-MURIATICUM. Nitro-muriatic Acid. (Acidi Nitrici, mensura, partem iij; Acidi Muriatici, mensura, partes v. Mix them in a vessel kept cool, and preserve the mixture in a well-stopped bottle, in a cool, obscure place.)

Stimulant, antiseptic.

Use. Largely diluted, it has been strongly recommended in malignant scarlatina, in chronic affections of the liver, and in syphilis; and still more diluted as a bath, in chronic derangement of the hepatic secretion, which it improves, and acts gently on the bowels.

Dose. ♏vi to ♏x, in f℥iij of water, twice or thrice a day. When used as a bath, the mixed acid should be added to the water until it tastes as sour as weak vinegar.

Incomp. Oxides, earths, alkalies, the sulphurets, and the acetates of potassa and of lead.

ACIDUM SULPHURICUM. Sulphuric Acid.

Escharotic, stimulant, rubefacient, tonic, astringent, refrigerant.

Use. In local pains, in the form of an ointment made of lard ℥j; sulphuric acid ʒj; and in scabies with ʒss. of the acid to the lard ℥j.

ACIDUM SULPHURICUM AROMATICUM. Aromatic Sulphuric Acid. Elixir of Vitriol.

Sulphuric acid 6 oz., Ginger 1 oz., Cinnamon 1½ oz.

Add the acid gradually to 1 pint Alcohol. Mix the Ginger and Cinnamon, and percolate with Alcohol to a pint of tincture. Then mix the dilute acid and tincture.

Use. In dyspepsia; the debility following intermittents, and other fevers, combined with vegetable bitters; and in chronic asthma.

Dose. ♏x to ♏xxx in fluids, twice or thrice a day.

ACIDUM SULPHURICUM DILUTUM. Diluted Sulphuric Acid. (Acidi Sulphurici f℥iss. Aquæ destillatæ f℥xivss. Mix gradually, and filter.)

Tonic, astringent, refrigerant.

Use. In dyspepsia, diabetes, menorrhagia, hæmoptysis, cutaneous eruptions, hectic; in gargles, in cynanche, and to check salivation. Sulphuric acid is an excellent tonic, and also possesses refrigerant and astringent properties, rendering it a valuable remedy in cases where we wish to avoid diarrhœa. In cases of low and hectic fever, attended with copious perspiration, it is very beneficial, as well as in hæmatemesis. It is also useful conjoined with saline aperients, when the urine has a tendency to phosphatic depositions, attended with loss of appetite, impaired digestion, foul tongue, etc. It is usually given with some bitter infusion, as cascarilla, columbo, cinchona, quassia, etc.

Dose. ♏x to ♏xx largely diluted; in gargles f℥j to f℥iij in ℥viij of fluid.

ACIDUM SULPHUROSUM. Sulphurous Acid.

Use. In parasitic skin diseases, in lotion with 2 or 3 measures of glycerin or water. The sulphite of soda better for internal use.

ACIDUM TANNICUM. Tannic Acid. (*Tannin.*)

Use. Tannic acid may be advantageously employed in all the passive hemorrhages, especially menorrhagia; also in diarrhœa, where we wish simply an astringent effect. It possesses a great advantage over most other astringents, from the smallness of dose in which it may be given, and from its being less liable to irritate the stomach and bowels.

Dose. From 2 to 4 grs. every three hours.

ACIDUM TARTARICUM. Tartaric Acid.
Refrigerant, antiseptic.
Use. In inflammatory affections, fevers, and scorbutus.
Dose. Gr. x to ℨss. dissolved in water.
Incomp. Alkalies and their carbonates, all the salts of potassa.

ACIDUM VALERIANICUM. Valerianic Acid.
Dose. Gtt. v-xv as a nervine.

ACONITIA.
Use. Externally counter-irritant; too poisonous to be used internally.

ACONITI FOLIA ET RADIX. Aconite. (*Aconitum Napellus.*)
Narcotic, sudorific, antiphlogistic.
Use. In chronic rheumatism, scrofula, scirrhus, palsy, amaurosis, and venereal nodes. Aconite is a very powerful topical remedy, in the form of tincture, in cases of rheumatism and neuralgia.
Dose. Gr. j, gradually increased to gr. v, twice or thrice a day; of the extract, from gr. ss. to gr. j, of the tincture from 10 to 40 drops, gradually increased.

ACTÆA. Baneberry. White and red Cohosh.
Root purgative and emetic.

ADANSONIA DIGITATA. Baobab.
Mucilaginous, diaphoretic. Used in miasmatic diseases of West Indies, in decoction of leaves and bark to 1 pint, to be taken in a day.

ADEPS. Lard.

Emollient.

Use. In ointments, etc.

ADIANTUM PEDATUM. Maidenhair.

An indigenous fern, bitter, aromatic, used in pectoral affections.

ÆSCULUS HIPPOCASTANUM. Horse-chestnut.

The bark has been substituted for cinchona; but it is uncertain.

ÆTHER SULPHURICUS. Sulphuric Ether.

Diffusible stimulant, narcotic, antispasmodic; externally refrigerant.

Pure washed sulphuric ether preferable to chloroform for anæsthetic purposes. It may be inhaled from a sponge.

Use. Hysteria, asthma, tetanus, epilepsy, and other spasmodic complaints; externally in headache, and dropped into the meatus in earache; it has also been used in burns.

Dose. ♏xx to f℥j in f℥xij water, or other fluid.

AGARIC. Spunk. Touchwood. Product of a fungus, the *Boletus*. The Boletus of the larch has been used as cathartic, that from the oak as styptic.

AGAVE AMERICANA. American aloe.

Diuretic and antisyphilitic.

Mexican *Pulque* is the fermented juice of this plant.

A species called *A. Virginica* has been used for bites of serpents.

AGRIMONIA EUPATORIA. Common Agrimony.

Astringent, in passive hemorrhages, etc.

Dose. ʒj or more in powder or infusion.

AILANTHUS GLANDULOSA.

Used as a vermifuge in tape-worm.

Dose. 8 to 30 grs. of powdered bark.

AJUGA CHAMÆPITYS. Ground pine.

Stimulant, balsamic, diuretic, aperient.

Dose. ℨj–ℨij of powdered leaves, or in an infusion of wine.

AKAZGA. Boundou. An African ordeal poison, similar to Nux Vomica.

ALCHEMILLA VULGARIS. Ladies' mantle.
Astringent. Used in diarrhœa.

ALCOHOL. Alcohol.
Stimulant (powerful and diffusible), sedative.

Use. Scarcely ever used internally in its pure state, but sometimes advantageously in a highly diluted form; in cases of debility and low fevers; externally as a fomentation in muscular pains; to burns; and to restrain hemorrhages. The use of alcohol as a medicine has been much diminished within the last ten years. It is found unsuited to a great majority of cases of disease, and when employed, too often inducing an artificial appetite not easily overcome. From its strong attraction for water, it causes thickening or scirrhus of the stomach, and an indurated state of the liver; and from its powerful effects upon the nervous system, it induces epilepsy, tremors, coma, mania, and death. For these reasons, and that we have useful substitutes, it should seldom be prescribed.

ALCOHOL AMYLICUM. Fusel oil.
An active irritant poison, obtained from too long continued distillation of grain, etc.

ALCOHOL DILUTUM. Dilute Alcohol. Spiritus tenuior.
Alcohol and water, equal parts.

ALCOHOL, METHYLIC. Spiritus Pyroxilicus. Wood Naphtha. Used in the arts, etc., as a cheap substitute for Alcohol. Has been considered narcotic and sedative in doses of 10 to 40 drops, diluted, 3 times a day.

ALCORNOQUE.
A bark from S. America. Astringent. Febrifuge.
Dose. 30 grs. of powder.

ALETRIS. Star-grass. (*A. farinosa.*)
An intense bitter, tonic, stomachic.
Use. In rheumatism and dropsy.
Dose. Of powder 10 grs., fluid extract 10 to 30 drops.

ALEURITES TRILOBA. A widely-diffused tropical plant, whose oil has been proposed as a substitute for *Castor oil.* It is known in Jamaica as Spanish walnut oil, and in the Sandwich Islands as Kukui oil.

ALISMA PLANTAGO. Water plantain.
Root has been used in epilepsy, chorea, and hydrophobia, in doses of powder from 10 grs. upwards.

ALKANET. Root of Anchusa tinctoria.
Used chiefly as a coloring material.

ALLIARIA OFFICINALIS. Hedge garlic.
Diuretic, diaphoretic, and expectorant.

ALLIUM. Garlic Bulbs. (*A. sativum.*)
Stimulant, diuretic, expectorant, emmenagogue, diaphoretic, and anthelmintic; extremely rubefacient, maturient, and repellent.

Use. In cold leucophlegmatic habits, dropsy, rheumatalgia, humoral asthma, and hysteria. Intermittents have been cured by it. The juice dropped into the ear, in atonic deafness, is a very effectual remedy; and it is also beneficial in herpetic eruptions, formed with oil into an ointment. A poultice of it over the pubis has been found useful in atony of the bladder.

Dose. One to six cloves, swallowed without chewing, twice or thrice a day. Of the juice f℥ss. to f℥ij mixed with sugar or syrup. In pills with soap or calomel, gr. xx to ℈ij.

ALNUS RUBRA. Tag Alder.

Alterative, emetic, astringent.

Use. In scrofula, secondary syphilis, etc.

Dose. Of fluid extract f℥j to f℥ij.

ALOE. Peculiar bitter principle (Aloin).

Cathartic, warm and stimulating, emmenagogue, anthelmintic, stomachic; hurtful in hemorrhoids. Aloes act chiefly on the large intestines, and produce catharsis by increasing the peristaltic muscular action, and not by increasing the secretions.

Dose. To act as a cathartic, gr. ij to gr. x; as an emmenagogue, gr. j to gr. ij, twice or thrice a day.

ALTHÆÆ FOLIA ET RADIX. Marshmallow Leaves and Root. (*A. officinalis.*)

Emollient, lubricating, demulcent.

Use. In pulmonary and intestinal affections; ardor urinæ; calculus; externally in fomentations, clyster, and gargles.

ALUMEN. Alum.

Astringent; and in large doses laxative, emetic.

Use. In hemorrhages, leucorrhœa, croup, hooping-cough, etc.; externally in relaxation of the uvula, ophthalmia, gleet, and fluor albus.

Dose. Gr. x to ℈j in powder, or in whey, made with ℨij of the powder, to Oj of hot milk, a teacupful occasionally; in gargles ℨss. in f℥iv of fluid; in collyria and injections gr. xij in f℥vj of rose water. A saturated solution is a useful styptic. Alum Curd is a good cooling external application in ophthalmia and other diseases; made by beating up the white of an egg with a piece of alum till it forms a coagulum.

Incomp. Potassa and potassæ carbonas, sodæ carbonas, ammonia, lime, magnesia, acetate of lead, infusion of galls.

ALUMEN EXSICCATUM. Dried Alum. (Melt the alum in an earthen vessel over the fire, until the ebullitions cease.)

Escharotic.

Use. To destroy fungus in ulcers; internally in colic.

Dose. Gr. iv to xij.

ALUMINÆ SULPHAS. Sulphate of Alumina.

Used externally, as astringent and antiseptic.

AMARANTHUS HYPOCHONDRIACUS. Prince's Feather.

Leaves astringent.

AMBERGRIS.

Antispasmodic.

Dose. 5 grs. to ʒj.

AMBROSIA TRIFIDA. Ragweed.

Astringent, tonic.

AMMONIACUM. Gum Ammoniac. (*Dorema ammoniacum.*)

Expectorant, antispasmodic, discutient, resolvent.

Use. In asthma and chronic catarrh; visceral obstructions, and obstinate colic from viscid matters lodged in the intestines; externally in scirrhous tumors and white swelling of the joints.

Dose. Gr. x to ℨss. in pills, with squill, myrrh, etc., or in emulsion. See Mist. Ammoniaci.

AMMONIÆ ARSENIAS. Arseniate of Ammonia.

Use. In inveterate skin disease.

Dose. 20 drops of solution of 1 gr. to f ℨj Dest. Water.

AMMONIÆ BENZOAS. Benzoate of Ammonia.

Diuretic.

Dose. Gr. xv to gr. xxx.

AMMONIÆ BICARBONAS. Bicarbonate of Ammonia.

Antacid. Similar to Bicarbonate of Soda, but more stimulant.

AMMONIÆ BORAS. Biborate of Ammonia.

Use. In chronic catarrh of bladder.

Dose. 10 to 20 grs. in water, often repeated.

AMMONIÆ CARBONAS. Sesquicarbonate of Ammonia.

Stimulant, antacid, diaphoretic, antispasmodic.

Use. In hysteria, dyspepsia, chronic rheumatism; applied to the nostrils in syncope.

Dose. Gr. v to ℈j in pills, or any bland fluid. Gr. xxx are an emetic.

Incomp. Acids, potassa fusa, liquor potassæ, magnesia, carbonates, alum, chloride of calcium, bitartras and bisulphas of potassæ, salts of iron, with the exception of the potassio-tartrate; bichloride of mercury, salts of lead, sulphate of zinc, sulphate of quinia.

AMMONIÆ MURIAS. Hydrochlorate of Ammonia. Sal Ammoniac.

Aperient, diuretic; externally to produce cold during its solution; stimulant.

Use. Externally while dissolving, to abate the heat and pain of inflammation; to allay headache; in lotion, composed of the salt ℨj, alcohol f℥j, water f℥ix, to indolent tumors, gangrene, scabies, and chilblains. Has been used internally in doses of grs. ij to grs. v, with Ext. Taraxaci and Rhei, as a substitute for Calomel in hepatic disease.

Dose. Gr. v to ʒss.

Incomp. Sulphuric and nitric acids, acetate of lead, potassa, carbonates of soda and potassa, lime.

AMMONIÆ PHOSPHAS. Phosphate of Ammonia.

Use. In gout and rheumatism.

Dose. Gr. x to xl, 3 or 4 times a day, in a tablespoonful of water.

AMMONIÆ VALERIANAS. Valerianate of Ammonia.

Use. In neuralgia, chorea, epilepsy, etc.

Dose. 2 to 8 grs., in water, or in pills.

AMMONII BROMIDUM. Bromide of Ammonium.
  Use. Similar to Bromide of potassium. Thought to influence the ganglionic functions.
  Dose. 2 grs. to 20 grs.

AMMONII IODIDUM. Iodide of Ammonium.
  Used externally as a substitute for Iodine.

AMPELOPSIS QUINQUEFOLIA. Virginia Creeper.
  Alterative, tonic, expectorant. Used in decoction or infusion.

AMYGDALÆ AMARÆ ET DULCIS. Bitter and Sweet Almonds. (*A. communis.*)
  Demulcent, the bitter is sedative.
  Use. In inflammatory complaints; and as a vehicle for more active remedies.

AMYLUM. Starch.
  Demulcent, nutritious.
  Use. In dysentery, tenesmus, and ulceration of the rectum, in the form of a clyster; it is the common vehicle for exhibiting opium per anum. The decoction of starch is made by boiling, for a short time, ʒiv starch in Oj water, previously mixing them gradually while the water is cold.

ANACARDIUM OCCIDENTALE. Cashew-nut.
  Juice used in the W. Indies for cure of corns, etc.

ANAGALLIS ARVENSIS. Scarlet Pimpernel.
  Used as preventive of hydrophobia, etc. Uncertain.

ANCHUSA OFFICINALIS. Bugloss.
  Used in France similarly to Borage.

ANDROMEDA ARBOREA. Sorrel-tree.
  Acid leaves form a refrigerant decoction in fevers.

ANEMONE PRATENSIS. Meadow Anemone.
  Used in ophthalmia and catarrhal inflammations.
  Dose. Of powder 2 or 3 grs. Tincture fʒss.
  A. pulsatilla is a favorite homœopathic remedy.

ANETHUM. Dill Seed. (*A. graveolens.*)
  Stimulant, carminative.

Use. In flatulent colic, and hiccough, particularly in infants.

Dose. Gr. x to ʒj.

ANGELICA.
Aromatic tonic.
Dose. Of root or seeds, 30 grs. to ʒj.

ANGUSTURA. Cusparia. (*Galipea officinalis.*)
Tonic, stimulant, aromatic.
Use. In dyspepsia, removing flatulence and acidity; chronic diarrhœa, dysentery.
Incomp. Sulphate of iron and of copper, nitrate of silver, tartarized antimony, acetate and diacetate of lead, bichloride of mercury, pure potassa, and infusions of galls and yellow cinchona bark, etc.
Dose. Gr. v to gr. xx in powder.

ANISUM. Aniseed. (*Pimpinella anisum.*)
Carminative.
Use. In dyspepsia and the tormina of infants.
Dose. Gr. x to ʒj bruised.

ANNOTTA.
Used principally as a coloring-matter.

ANTENNARIA MARGARITACEA. Life-everlasting.
Leaves astringent and expectorant.

ANTHEMIS. Chamomile flowers. (*A. nobilis.*)
Tonic, stomachic; the warm infusion is emetic; externally discutient, emollient, antiperiodic.
Use. In intermittents, dyspepsia, hysteria, flatulent colic, gout; to promote the operation of emetics, externally as fomentations in gripings, and to ripen suppurating tumors.
Dose. In powder ʒss. to ʒij twice or thrice a day.

ANTIMONII ET POTASSÆ TARTRAS. Potassio-Tartrate of Antimony, or Emetic Tartar.
Emetic, sometimes cathartic, diaphoretic, expectorant, alterative, rubefacient. A sedative to the circulation, while it increases most of its secretions.

Use. In the beginning of fever, to clear the stomach and bowels; but it is an improper emetic in advanced stages of typhus; in large doses in pulmonary inflammations; and in small as an alterative in cutaneous diseases, acute rheumatism, chorea; externally in white swellings, hooping-cough, phthisis, and all deep-seated inflammations.

Dose. As a means of subduing inflammation, gr. ss. to gr. ij; as an emetic, gr. j to gr. iv, in solution; diaphoretic and expectorant, gr. 1-12th to 1-8th. It is made into an ointment for external use, by rubbing up ℥ij with lard ℥j.

Incomp. Alkalies and earths with their carbonates; strong acids; hydro-sulphurets; lime-water, chloride of calcium, salts of lead; decoctions of bitter and astringent plants.

ANTIMONIUM SULPHURATUM.
Precipitated Sulphuret of Antimony.
Alterative, diaphoretic, emetic.
Use. In secondary syphilis, cutaneous eruptions, etc.
Dose. Gr. j to ij as alterative.

ANTIRRHINUM LINARIA. Toad-flax.
Diuretic, cathartic, slightly acrid.
Used in infusions, or as a poultice, etc., to piles.

APOCYNUM ANDROSÆMIFOLIUM. Dog's Bane.
Emetic, diaphoretic, alterative.
Dose. Gr. xxx of the powdered root as an emetic; gr. v diaphoretic. Employed by the Indians in lues venerea.

APOCYNUM CANNABINUM. Indian Hemp.
Emetic, hydragogue, cathartic, diuretic, diaphoretic, expectorant, narcotic, and sedative.
Use. A very powerful remedy in ascites and general dropsy; in neuralgia, etc.
Dose. From gr x to gr. xx of the powdered root

produce free vomiting and purging. Of the decoction, which is preferable, and made by boiling ℥ss. of the dried root in Ojss. of water to Oj, from f℥j to f℥ij may be given three or four times a day, if necessary. Of the extract, gr. iij to gr. iv, two or three times a day, will usually act on the bowels.

AQUA AMMONIÆ. Solution of Ammonia.
 Stimulant. Antacid. Rubefacient.
 Dose. 10–30 drops, largely diluted.

AQUÆ. Medicated Waters. See p. 155.

AQUILEGIA VULGARIS. Columbine.
 Diuretic, diaphoretic, antiscorbutic, seldom used.

ARALIA NUDICAULIS. False Sarsaparilla.
 Stimulant, diaphoretic, alterative.
 Use. Employed in rheumatism, syphilis, cutaneous affections, in the same manner and dose as the genuine sarsaparilla.

ARALIA SPINOSA. Angelica-tree, Toothache-tree, Prickly Ash.
 Stimulant, diaphoretic, emetic, cathartic.
 Use. Employed in chronic rheumatism and cutaneous eruptions. Also, in colic, in toothache, usually given in decoction.

ARCTIUM LAPPA. Burdock.
 Aperient, sudorific, diuretic.
 Use. In rheumatism, gout, aphthæ, also in venereal, scorbutic, scrofulous, and nephritic affections; in decoction made with ℥ij of the root in Ojss. of water. The leaves externally in cutaneous eruptions and ulcerations.
 Dose. A teacupful several times a day; of little value unless persevered in for a long time.

ARECA NUT. Betel Nut.
 Used in tape-worm in doses of ʒiv to ʒvi.

ARGEMONE MEXICANA. Prickly Poppy.
 Anodyne cathartic.

Dose. 8 grs. of seeds in emulsion.

ARGENTI CYANIDUM. Cyanuret of Silver. Cyanide of Silver.

Use. To prepare hydrocyanic acid.

ARGENTI IODIDUM. Iodide of Silver. (Precipitated from the nitrate by iodide of potassium.)

A substitute for nitrate of silver in gastric affections, in doses of ⅛ gr. three times a day, increased gradually. This iodide forms a crystalline salt with iodide of potassium, which may perhaps combine the tonic and alterative effects of its constituents, in a similar way to the iodo-hydrargyrate of potassium. It is, however, decomposed by water.

ARGENTI NITRAS. Nitrate of Silver.

Tonic, antispasmodic, escharotic.

Use. In chorea, epilepsy, dyspepsia, and irritable conditions of the mucous membrane of the stomach and bowels; locally to relieve strictures; to fungous ulcers, warts, and venereal chancres; gr. ij in distilled water f℥j is a good injection in fistulous sores; and as an application to spongy gums, enlarged tonsils and ulcerated sore throats. A solution of ℨss. in f℥j of distilled water, is highly useful when pencilled over the surface in erysipelas.

Dose. Gr. ⅛ increased to gr. iv in a pill, with mucilage, three times a day; or in solution, increased to gr. iij. The dark color communicated to the skin is an objection to its internal employment.

Incomp. Sulphuric, hydrochloric, and arsenious acids and their salts; alkalies except ammonia; lime; chlorides; sulphurets; astringent vegetable infusions and decoctions; aqueous solutions of salts of mercury, or of copper.

ARGENTI NITRAS FUSUS. Lunar caustic.

Fused nitrate of silver.

Convenient for external use, as stimulant or escharotic.

ARGENTI OXIDUM. Oxide of Silver.
Proposed as a substitute for the nitrate.
Dose. ½ a gr. in pill.

ARMORACIA. Horseradish Root. (*Cochlearia armoracia.*)
Stimulant, diuretic, diaphoretic.
Use. In scorbutus, rheumatism, dropsy, and dyspeptic affections; and locally in hoarseness.
Dose. ℈j to ʒj. Vide Infusion: of the following syrup a teaspoonful often slowly swallowed, in hoarseness: (℞. Of the scraped root ʒj, boiling water ℥ij, sugar q. s. to the strained liquor.)

ARNICA. Leopard's Bane. (*A. montana.*)
Narcotic, stimulant, diaphoretic, emmenagogue, diuretic.
Use. In amaurosis, paralysis, rheumatism, gout, dropsy, nephritis, and chlorosis. The root has been used in intermittents, but is most useful in diseases attended with a typhoid state of the system.
Dose. Gr. v to gr. x in powder, or f℥ss. of the following infusion: (℞. Of the flowers ʒjss., water f℥viij), twice or thrice a day. In large doses it produces poisoning.

ARUM. Dragon Root, Indian Turnip. (*A. triphyllum.*)
Externally irritant. Internally stimulant to all the secretions, especially those of the skin and lungs.
Use. In asthma, pertussis, chronic catarrh, chronic rheumatism, and cachectic complaints generally.
Dose. Of the powder of the recently dried root, gr. v to gr x, mixed with gum Arabic, sugar, and water, in the form of an emulsion, repeated two or three times a day, and gradually increased to ʒss. or ʒj. Also in aphthous sore mouth of children, mixed with sugar and laid on the tongue.

ASARABACCA. *A. Europæum.*

Emetic, cathartic, errhine.

Dose. Grs. 30-ʒj of powder.

ASARUM CANADENSE. Wild Ginger.

A stimulant, tonic, diaphoretic.

Use. Used chiefly as an elegant adjunct to tonic infusions and decoctions. Resembles serpentaria in its effects.

Dose. Of the powder, from gr. xx to gr. xxx. Also, used in form of a tincture.

ASCLEPIAS. The Common Silk-weed. Butterfly-weed. Pleurisy Root. (*A. tuberosa; Syriaca,* etc.)

Diaphoretic, expectorant, cathartic, diuretic, slightly tonic.

Use. Employed extensively in some of the Southern States in catarrh, pneumonia, pleurisy, consumption, acute rheumatism, autumnal remittents, and dysentery.

Dose. Of the powder, gr. xx to ʒj, several times a day. As a diaphoretic, a teacupful of the decoction every three or four hours, made by infusing ʒj of the roots in Oij of water.

ASCLEPIAS CURASSAVICA. Bastard Ipecacuanha.

Emetic, cathartic, astringent.

Dose. Ɉj to Ɉij. Expressed juice ʒj.

ASPARAGUS OFFICINALIS. Asparagus.

Diuretic, aperient.

Dose. ʒss. to ʒj of extract, or ʒj to f℥ij of syrup, prepared from the shoots.

ASSAFŒTIDA. Assafœtida. (*Narthex assafœtida.*)

Antispasmodic, expectorant; emmenagogue; anthelmintic when injected into the rectum.

Use. Hysteria, tympanitis, asthma, dyspnœa, pertussis, worms.

Dose. In pill, gr. x to ʒss.: in solution, vide *Mistura;* in clyster, ʒij dissolved in water f℥viij.

ASTER PUNICEUS.
: Stimulant, diaphoretic. Used in rheumatism and catarrh.

ATOMIZERS. By using medicated spray, many remedies may be brought in direct contact with the respiratory passages. To ℥j water use 10–30 grs. Alum; 10–20 grs. Muriate Ammonia; 3–10 ♏ Fluid ext. Hyoscyamus; 2–10 ♏ Tinct. Opii; 1–10 grs. Nit. Silver; 1–20 grs. Tannin, etc.

ATROPIA. Active principle of Belladonna.
: *Use.* One drop of a solution of gr. j to ℥iv of distilled water with a few drops of acetic acid, applied to the inner surface of eyelid, dilates the pupil in a few minutes.

Gr. j to ʒj of lard as an ointment in neuralgia.

ATROPIÆ SULPHAS. Sulphate of Atropia.
: Use same as Atropia, but more soluble.

AURANTII CORTEX. Orange-Peel.
: *Use.* A mild tonic, carminative, and stomachic.

AVENÆ FARINA. Oatmeal.
: Oatmeal gruel is made by boiling ℥j in Oiij water to a quart (with a few raisins). Strain, cool, and pour off the clear liquor. Add sugar and lemon-juice.

AZEDARACH. Bark of root *Melia Azedarach.*
: *Use.* Cathartic, emetic, anthelmintic — in large doses narcotic.
: *Dose.* Oii water to ℥iv bark, boiled to Oj; to a child a tablespoonful.

BALSAMUM PERUVIANUM. Peruvian Balsam. (*Myrospermum peruiferum.*)
: Stimulant, tonic, expectorant.
: *Use.* In palsy; chronic asthma, bronchitis, and rheumatism; gleet; leucorrhœa: and externally for cleansing and stimulating foul, indolent ulcers.

Dose. ♏x to ʒss. twice or thrice a day, made into an emulsion with mucilage of gum.

BALSAMUM TOLUTANUM. Tolu Balsam. (*Myrospermum toluiferum.*)
Stimulant, expectorant.
Use. In chronic coughs; but principally used on account of its flavor.
Dose. Gr. x to ʒss. triturated with mucilage.

BAPTISIA TINCTORIA. Wild Indigo.
Antiseptic, laxative, emetic.
Use. In scarlatina, gangrene, etc.
Dose. f℥ss. of decoction (℥j root to Oj water).

BEBEERU. Bark of *Nectandra Rodiei.*
The *sulphate of bebeerin* has been used in intermittent and remittent fevers.
Dose. 2 to 5 grs.

BELÆ FRUCTUS. Bael fruit. (*Ægle marmelos.*)
Astringent in diseases of the bowels.
Dose. Of decoction ℥ij to Oj, boiled to f℥iv, f℥j–f℥ij. Of extract fʒj to fʒij.

BELLADONNA. Deadly Nightshade. (*Atropa belladonna.*)
Powerful narcotic, diaphoretic, diuretic, repellent.
Use. In obstinate intermittents, tic-douloureux, palsy, epilepsy, chorea, mania, gout, rheumatism, dropsy, jaundice, pertussis, and the cachexiæ; amaurosis; sprinkling the powdered leaves over cancerous sores has been found to allay the pain; and the leaves form a good poultice. Applied to the eye, in the form of infusion or solution of the extract, to dilate the pupil previous to operations. The root is used for the same purpose as the leaves.
Dose. Gr. ss. gradually increased to gr. j daily; or f℥ij of this infusion: ℞. Of the leaves ℈j, hot water f℥x, strained cold.

BENZOIN ODORIFERUM. Spice-bush. Fever-bush.

Stimulant, aromatic, vermifuge.
Used in decoction or infusion.

BENZOINUM. Benzoin. (*Styrax Benzoin.*)
Use. Principally for obtaining the acid it contains.
Dose. Gr. x to ℨss.
Incomp. Alkalies, acids; and so with all the balsams.

BENZOLE. Benzin. (*Hydruret of Phenyle.*) A constituent of coal-gas tar.
Use. Vapor is anæsthetic. With four parts of lard in scabies, etc. As a liniment in rheumatism and neuralgia Used internally in trichiniasis.

BERBERIS VULGARIS. Barberry.
Refrigerant, astringent, antiscorbutic.
The berries are used as a drink, and the juice as syrup.

BETULA LENTA. Sweet Birch. Black Birch.
Gently stimulant and diaphoretic.

BISMUTHI SUBCARBONAS. Subcarbonate of Bismuth.
Used as substitute for subnitrate.
Dose. 15–45 grs., in water, before meals.

BISMUTHI SUBNITRAS. Subnitrate of Bismuth.
Antispasmodic. Absorbent. Sedative.
Dose. 5–15 grs. in pill, etc.

BISMUTHI VALERIANAS. Valerianate of Bismuth.
Use. In neuralgia and gastralgia.
Dose. Gr. ss. to gr. ij in pill.

BISTORT. *Polygonum bistorta.*
Mild astringent.
Dose. 20–30 grs. of powder.

BISULPHATE OF POTASSA.
Aperient tonic.
Dose. ℨj to ℨij.

BISULPHURET OF CARBON.
Diffusible stimulant. Vapor anæsthetic.

BITTERA FEBRIFUGA. Bitter Ash.
Similar to Quassia.

BOLE ARMENIAN.
An argillaceous earth. Used chiefly in tooth-powders.

BORAGO OFFICINALIS. Borage.
Mucilaginous, gently diaphoretic.
Dose. ℨij to ℨiv of expressed juice.

BROMINIUM. Bromine.
Like iodine, it stimulates the lymphatics, and promotes absorption.
Use. In bronchocele, scrofula, etc.
Dose. In aqueous solutions (1 part to 40 of water), 6 drops several times a day.

BRYONY. *B. alba.*
Hydragogue, cathartic.
Dose. ℈j to ʒj of powdered root.

BUCHU. Diosma. Leaves of *Barosma crenata.*
Stimulant, diuretic.
Use. In irritation of the bladder.
Dose. 20 to 30 grs. powder, f ℨj to f ℨij of infusion (ℨj to Oj boiling water), ʒj of the fluid extract.

BUXUS SEMPERVIRENS. Box.
Diaphoretic, purgative.
Dose. ʒj of leaves as a purge.

CABBAGE-TREE BARK. *Andira inermis.*
Cathartic vermifuge.
Dose. Powder ℈j to ʒss. Extract 3 grs. Decoction f ℨij.

CACTUS GRANDIFLORA. Night-blooming Cereus.
A tincture of fresh stems macerated a month in alcohol, used in functional palpitation of heart.
Dose. 1 to 5 drops three times a day.

CADMII IODIDUM. Iodide of Cadmium.
Used in ointment, 1 part to 8 of lard, for enlarged glands, etc. A substitute for the iodide of lead.

CADMII SULPHAS. Sulphate of Cadmium.
 Resembles sulphate of zinc as astringent and emetic. Used also in syphilis, rheumatism, and opacities of cornea.

CAFFEA. Coffee. (*C. Arabica.*)
 Nervous stimulant.
 Dose. A cupful or more of infusion of ℨj to Oj boiling water.

CAHINCA. *Chiococca racemosa.*
 Tonic, diuretic, purgative, emetic.
 Dose. Powder ℈j to ʒj. Extract 10 to 20 grs.

CALAMINA. Calamine. Impure Carbonate of Zinc.

CALAMINA PRÆPARATA. Prepared Calamine.
 Use. A mild astringent in excoriations.

CALAMUS AROMATICUS. Sweet Flag-root.
 Stomachic, carminative.
 Use. In atonic dyspepsia with vertigo.
 Dose. ℈j to ʒj in powder.

CALCIS CARBONAS PRÆCIPITATA. Precipitated Carbonate of Lime.
 Use the same as Creta Præparata.

CALCIS PHOSPHAS PRÆCIPITATA. Phosphate of Lime.
 Use. In scrofula, rickets, etc.
 Dose. 10 to 30 grs.

CALENDULA OFFICINALIS. Marigold.
 Antispasmodic, sudorific, emmenagogue.
 Used in infusion.

CALLITRICHE VERNA. Water Star-wort.
 Diuretic. Given in decoction.

CALOTROPIS GIGANTEA. *Asclepias gigantea.*
 An East India remedy for cutaneous disease, syphilis, etc.
 Dose. 3 to 12 grs. three times a day.

CALUMBA. Columbo.

Mild tonic. Used in dyspepsia, etc.

Dose. 10 grs. to ℈j. In flatulence, etc., an infusion of ℥ss. Calumba, ℥ss. Ginger, ʒj Senna, and Oj boiling water. Give a wineglassful three times a day.

CALX CHLORINATA. Chlorinated Lime. Chloride of Lime. Pass chlorine through lime until it is saturated.

Use. As a disinfectant, and for bleaching purposes.

CAMPHORA. Camphor. (*C. officinarum.*)

Narcotic, diaphoretic, sedative; externally anodyne.

Use. In typhus, cynanche maligna, confluent small-pox, and other exanthemata of the typhoid type; in atonic gout, and as an adjunct to bark and opium in checking gangrene. It produces its narcotic and sedative effects with very little increase of pulse, and therefore may be used in mania, pneumonia, and other inflammatory complaints, united with nitre and antimonials. In doses of from 1 to 3 grains it acts as a diaphoretic. It is a useful adjunct to bark in typhoid diseases, to valerian, the fetid gums, volatile alkali, and others, in hysteric and nervous complaints, and to antimonials in rheumatism and other inflammations. Externally, it allays the pains of rheumatism and other deep-seated inflammations, when dissolved in oil.

Dose. Gr. ij to gr. x in powder, with sugar, etc.; in pills; or in mixture with mucilage, or almond confection; the effects of an overdose are counteracted by opium. For external application it is dissolved in oil or in alcohol.

CANELLA. Canella Bark. (*C. alba.*)

Stimulant, tonic.

Use. As an aromatic addition to bitter tonics and cathartics.

Dose. Gr. x to ʒss. in powder; in infusion f℥iss.

CANTHARIS. The Blistering Fly.

Stimulant, diuretic, rubefacient, vesicant; both their internal use and their external application are apt to produce strangury; active properties depend on the cantharidin.

Use. Internally in dropsies, obstinate gleet, and leucorrhœa; retention of the urine owing to want of action in the bladder, and incontinence of urine from debility of the bladder; lepra; but their internal use requires caution. For their external use, see Empl., Tinctura, and Acetum Cantharidis.

Dose. Gr. ss. to gr. j, in a pill, with opium or the extract of henbane and camphor, twice a day.

CANTHARIS VITTATA. Potato Fly.

Same as the former. There are several other species, all of which have the same properties.

CAPSICUM. The Capsicum berries. (*C. annuum.*)

Stimulant, rubefacient.

Use. In atonic gout, the flatulence of dyspepsia, lethargy. Its solution (Capsici pulv. ʒj, Sodii Chlor. Ɔj, Acet. ʒiv, Aquæ ferventis f ʒvj, Cola) forms the best gargle in cynanche maligna, and scarlatina. Cataplasms of it are used in coma and the delirium of typhus.

Dose. Gr. iij to gr. x in pill, or f ʒss. of infusion.

Incomp. Nitrate of silver, bichloride of mercury, acetate of lead, sulphates of iron, zinc, and copper, and the carbonates of alkalies.

CARBO ANIMALIS. Animal Charcoal. (Prepared from flesh and bones.)

Use. For decolorizing vegetable salts, clarifying salts, and extracting the volatile oil from whiskey and other liquors.

CARBO LIGNI. Charcoal of Wood.

Antiseptic, absorbent.

Use. In putrid eructations of dyspepsia, obstinate

constipation; to relieve the nausea of pregnancy, and as a cataplasm with linseed meal to fetid ulcers; the best tooth-powder.

Dose. Gr. x to ℈j united with rhubarb.

CARDAMINE PRATENSIS. Cuckoo-flower.
Antispasmodic, diuretic.

CARDAMOMUM. Cardamom Seed. (*Elettaria cardamomum.*)
Carminative, stomachic.

Use. In the flatulent colic of children, united with rhubarb and magnesia; but principally to give warmth to other remedies.

Dose. Gr. v to ʒj in powder.

CAROTA. Carrot Seed.
Mild diuretic.

Dose. 30 grs. to ʒj of bruised seeds.

CARTHAMUS. Dyer's Saffron. (*C. tinctorius.*)
Laxative and somewhat diaphoretic.

Use. As a substitute for saffron in measles, scarlatina, and other exanthematous diseases, to promote the eruption.

Dose. Of an infusion of ʒij to a pint of boiling water, given without restriction as to quantity.

CARUM. Caraway Seeds. (*C. carui.*)
Carminative.

Use. In flatulent colic, and to give warmth to purgatives.

Dose. Gr. x to ʒj, swallowed whole, or in powder.

CARYA. Hickory.
Astringent. Infusion of inner bark in dyspepsia and intermittents.

CARYOPHYLLUS. The Clove. (*C. aromaticus.*)
Stimulant, aromatic.

Use. As a corrigent to other remedies, and a condiment.

Dose. Gr. v to x, in powder.

CASCARILLA. Cascarilla Bark. (*Croton Eleuteria.*)
Tonic, stomachic.
Use. As an adjunct to chinchona in ague; in obstinate diarrhœa, and after dysentery; a good vehicle for powdered Peruvian bark, and small doses sulphate magnesia and sulphuric acid, in debility of stomach attended with constipation; in dyspepsia, and flatulent colic.
Dose. Gr. x to ʒj in powder. The infusion is the best form.

CASSIA. Cassia Pulp. (*C. fistula.*)
Laxative.
Use. Where a gentle medicine is required in costive habits, combined with aromatics.
Dose. ʒi to ʒj.

CASTANEA. Chinquapin. (*C. pumilla.*)
Tonic and astringent. Leaves of chestnut (*C. vesca*) similar.
Use. In intermittents.

CASTOREUM. Castor.
Antispasmodic, emmenagogue.
Use. In typhus, hysteria, epilepsy, amenorrhœa.
Dose. Gr. x to ϶j in a bolus; ʒj, or more, in clysters; of little value as a remedy.

CATALPA CORDIFOLIA. Catawba Tree.
Reputed poisonous, but has been used in asthma, ʒiij or ʒiv of seeds in ʒxij water, boiled to ʒvj, given morning and night.

CATARIA. Catnip. Leaves of *Nepeta Cataria.*
Tonic and excitant.
Use. In domestic practice in amenorrhœa, colic, etc.
Dose. ʒij in infusion.

CATECHU. Catechu. (*Acacia catechu.*)
Astringent, tonic.
Use. In diarrhœa, from a relaxed state of the

bowels, and in intestinal and uterine hemorrhages; locally in aphthæ, ulceration of the gums, and in coughs and hoarseness from the relaxation of the uvula, and epistaxis.

Dose. Gr. x to ʒj in powder; in the latter case, a piece is allowed to dissolve slowly in the mouth; but is best given with sugar, gum Arabic and water.

CAULOPHYLLUM THALICTROIDES. Blue Cohosh.

Emmenagogue, diaphoretic.

Dose. f ʒj to f ʒij of infusion or decoction, f ʒj to f ʒij of tincture.

CEANOTHUS AMERICANUS. New Jersey Tea.

Astringent. Used in syphilis in decoction (ʒij to Oj). A strong infusion used in aphthæ and dysentery.

CEDRON. *Sinaba cedron.*

Used in Central America as antidote of the bite of serpents, in hydrophobia, and intermittents.

Dose. 1 to 2 grs., or more.

CELASTRUS SCANDENS. Climbing Staff-tree.

Emetic, diaphoretic, narcotic.

CENTAUREA BENEDICTA. Blessed Thistle. May be used as a tonic in cold infusion (ʒss. of leaves to Oj). A stronger infusion taken warm in bed promotes diaphoresis. A still stronger infusion is emetic.

CENTAURY. *Erythræa centaurium.*

Tonic, like gentian.

Dose. 30 grs. to ʒj.

CEPHALANTHUS OCCIDENTALIS. Button-bush.

Laxative and tonic.

CERA FLAVA ET CERA ALBA. Yellow Wax and White.

Demulcent, emollient.

Use. In diarrhœa and dysentery; but principally used in the formation of cerates and ointments.

Dose. ƎJ to ʒss. twice or thrice a day, in form of emulsion: melt the wax with a little oil, then triturate it with yolk of egg, and groat gruel fʒij.

CERATA. Cerates.
See Pharmaceutical preparation, p. 175.

CERII OXALAS. Oxalate of Cerium.
Nervous tonic, in gastralgia, etc.
Dose. 1–2 grs., in pill or solution.

CETACEUM. Spermaceti.
Demulcent, emollient.
Use. In coughs and dysentery; and in the composition of ointments.
Dose. ʒss. to ʒjss., rubbed up with sugar, or with an egg, in emulsion.

CETRARIA. Iceland Moss. (*C. islandica.*)
Tonic, demulcent, nutrient. See Decoct. Cetrar.

CHELIDONIUM MAJUS. Celandine.
Acrid, purgative, diuretic, diaphoretic. Used in jaundice.
Dose. Gr. xxx, dried herb or root.

CHELONE GLABRA. Balmony.
Tonic, cathartic, anthelmintic.
Dose. ʒj fluid extract.

CHENOPODIUM. Wormseed. (*C. anthelminticum.*)
Anthelmintic.
Use. To destroy lumbrici in children, for which it is given morning and evening for three or four days; then followed by calomel, or some brisk cathartic.
Dose. Of the powdered seeds, from Ǝj to Ǝij to a child two or three years old, in syrup; of the oil, which is more frequently given, from 5 to 10 drops, mixed with sugar or in emulsion; of the decoction, made by boiling ʒj of the fresh plant in Oj of milk, with the addition of orange-peel, or some other aromatic, a wineglassful, or a tablespoonful of the expressed juice of the leaves.

CHIMAPHILA. Pipsissewa. Leaves of *Chimaphila Umbellata.*
Diuretic, tonic, astringent.
Use. In urinary affections, scrofula, and rheumatism.
Dose. Oj of decoction in 24 hours.

CHIRETTA. *Agathotes Chirayta.*
Similar to gentian.
Dose. Of powder 20 grs.

CHLORAL. Chloral Hydrate. Hydrate of Chloral. Prepared from the action of Chlorine on Alcohol. Hypnotic and anodyne.
Use. In delirium tremens and nervous diseases, to produce sleep.
Dose. Grs. x to ʒj in syrup, etc.

CHLOROFORM. Anæsthetic. From the accidents which have occurred from its use, pure Sulphuric Ether, or one part chloroform, two parts ether, is preferable for the purpose of procuring insensibility to the pain of surgical operations.
Dose. For internal use ♏x to fʒj.

CHONDRUS. Irish Moss, Carrageen. A good substitute for the Iceland Moss, which it closely resembles. (Macerate ℥ss. of the moss ten minutes in cold water, turn it off, add Ojss. boiling water; boil to a pint, strain, and add sugar and lemon-juice to improve the flavor. Milk may be substituted for water, if a more nutritious preparation be required.)

CICHORIUM INTYBUS. Chicory.
Tonic, aperient. Used in jaundice and phthisis, in decoction (℥j or ℥ij to Oj).

CICUTA VIROSA. Water hemlock.
Acrid narcotic poison.
Used externally in poultices or extract.

CIMICIFUGA. Black Snakeroot. Cohosh. (*C. racemosa.*)

Tonic, diuretic, diaphoretic, expectorant, emmenagogue.

Use. Employed extensively in the United States, as a domestic remedy, in rheumatism, dropsy, chorea, hysteria, and especially in pulmonary affections, for which it has been regarded by some as a specific.

Dose. Of the powder, gr. x to ʒj; of the decoction, made by boiling ʒj of the bruised root in Oj of water, from fʒj to fʒij may be given several times a day.

CINCHONA. Peruvian Bark. (*C. flava — pallida — rubra.*)

Strongly and permanently tonic and antiperiodic, slightly astringent, stomachic, and febrifuge.

Use. In intermittents, after evacuating the stomach and bowels; in continued fevers; keeping the bowels clear; confluent small-pox; erysipelas; acute rheumatism; cynanche maligna; scarlatina; passive hemorrhages; and in every disease attended with deficient action. Externally in clysters, gargles, and lotions, in gangrenous ulcerations, etc. To check the nausea excited by it, wine, aromatics, and carbonic acid are added; to prevent purging, opium; costiveness, rhubarb.

Dose. Given in infusion, decoction, and extract. The latter is a good form, if well prepared; of this gr. iij to gr. x in pill, or dissolved in infusion of roses, or syrup of orange-peel, three times a day. Of the powder, ℈j to ʒiij in infusion of liquorice, or water.

CINCHONIÆ SULPHAS. Sulphate of Cinchonia.
Similar to sulphate of quinia.
Dose. 1 gr. to 15.

CINNAMOMUM. Cinnamon Bark. Cassia.
Stimulant, astringent, carminative, tonic.
Use. As a grateful aromatic in dyspepsia and diar-

rhœa; to cover the taste of nauseous remedies, and with cathartics to prevent griping. The infusion checks vomiting. Chewed in palsy of the tongue.

Dose. Gr. v to ⴺj in powder.

CITRATE OF IRON AND MAGNESIA.
Saline chalybeate.
Dose. 5 to 10 grs.

CITRATE OF SODA.
Similar to Citrate of Magnesia.
Dose. In diabetes ʒss. to ʒj.

CLEMATIS ERECTA. Upright Virgin's-bower.
Diuretic and diaphoretic. Used in Bright's disease.
Dose. f℥iv of infusion of ʒiij of leaves to Oj water.

COBWEB. Spider's web.
Said to be antiperiodic, etc.
Dose. 5 or 6 grs. in pill.

COCCULUS INDICUS.
Acrid narcotic poison. Used externally in tinea capitis, etc.

COCCUS. The Cochineal Insect. The dried female.
Use. Chiefly for giving a red color to tinctures, etc. In doses of $\frac{1}{3}$ gr. in Hooping-cough.

COCHLEARIA OFFICINALIS. Scurvy-grass.
Stimulant, aperient, diuretic.

COCOS BUTYRACEA. The plant which affords *palm oil,* or *coco butter.*
This latter is of the consistence of an ointment, and is used as an excipient for suppositories and medicated pessaries.

COFFEE. *C. Arabica.*
Medically is excitant to the nerves, and resists the intoxicating effects of alcohol and opium.

COLCHICI RADIX ET SEMEN. The Bulb and Seeds of the Meadow Saffron. (*Colchicum autumnale.*)
Narcotic, diuretic, cathartic.
Use. In dropsies, gout, rheumatism, neuralgia,

bronchitis, and scarlet fever. Colchicum is rather palliative than curative in gout and rheumatism. It is a useful addition to saline medicines in fevers and all inflammations. It should be given in small doses combined with magnesia, and, if necessary, often repeated.

Dose. Gr. j to gr. v. of the recent bulb in pill.

COLLINSONIA CANADENSIS. Horse - weed. Heal-all. A decoction of fresh root used in catarrh of the bladder, leucorrhœa, dropsy, etc.; and a poultice of leaves to bruises and the like.

COLLODIUM. Collodion (a solution of gun-cotton in ether); an artificial epidermis.

Cantharidal collodion, for blisters, is prepared by the addition of an ethereal solution of cantharides.

COLOCYNTHIS. Bitter Cucumber Pulp. (*Cucumis colocynthis.*)

Strongly cathartic, drastic, hydragogue.

Use. Too violent to be used alone. When combined with calomel, extract of jalap, and gamboge, colocynth forms a highly efficient and safe cathartic, especially adapted to congestion of the liver and portal circle, and torpidity of this organ. In dropsy, and affections of the head, also highly useful.

Dose. Gr. j to gr. v.

COMPTONIA ASPLENIFOLIA. Sweet Fern.

Tonic, astringent. A decoction used in diarrhœa.

CONFECTIONES. Confections.

See Pharmaceutical preparations, p. 168.

CONIUM. Hemlock Leaves and Seeds. (*C. Maculatum.*)

Narcotic, poisonous in an overdose; resolvent.

Use. As a palliative in cancer and scirrhus, scrofulous, and syphilitic ulcerations and swellings; pertussis; chronic enlargements of the liver and other abdominal organs; cutaneous affections; asthma;

chronic pulmonary diseases, and neuralgic affections. Externally ℨiij of the dried herb boiled in Oj of water as a fomentation to open scrofulous and cancerous ulcers; or as a cataplasm, by adding linseed meal and oatmeal.

Dose. Gr. ij to gr. iv of the powder, or from gtt. ij of the expressed juice, very gradually increased to ℨj. Of the extract, gr. j to gr. iv; to be reduced if it causes vertigo. The extract is the best form of administering it; it may be usefully combined with ipecacuanha in pulmonary affections, where we wish to quiet cough and relieve bronchial irritation.

CONTRAYERVA.   Contrajerva Root.   (*Dorstenia contrayerva.*)

Tonic, stimulant, sudorific.

Use. In typhus, nervous fever; the fever of dentition in weak infants; and dysentery.

Dose. Gr. x to ℨss.

CONVALLARIA MAJALIS.   Lily of the Valley.

Powdered leaves, sternutatory.

CONVALLARIA MULTIFLORA.   Solomon's Seal.

Tonic, mucilaginous, mildly astringent. Used in leucorrhœa, etc.

Dose. Fluid extract ℨij to ℨv.

CONVOLVULUS PANDURATUS.   Wild Potato.

Feebly cathartic, diuretic.

Dose. 40 grs. of dried root.

COPAIBA.   Copaiba Balsam.   (*C. officinalis.*)

Stimulant, diuretic, purgative in large doses; acts on the urethra.

Use. In gonorrhœa, gleet, leucorrhœa, dysentery, and all affections of mucous membranes; hemorrhoidal affections.

Dose. ♏x to fℨj in emulsion with gum or yolk of egg; in pills, by mixing the copaiba with magnesia and exposing the mixture to the air. It is given also in gelatinous capsules.

Incomp. Sulphuric acid, nitric acid.

COPTIS. Goldthread. (*C. trifolia.*)
Tonic.
Use. In all cases where a simple tonic is required. In aphthous affections of the mouth and throat.
Dose. Of the powder from gr. x to gr. xxx. Of the tinct. ʒj, ʒj root, Oj alcohol.

CORALLORHIZA ODONTORHIZA. Coral-root.
Stimulant, diaphoretic.
Dose. 30 grs. of powder.

CORIANDRUM. Coriander Seed. (*C. sativum.*)
Carminative.
Use. In flatulencies; but chiefly to cover the taste of other medicines.
Dose. ℈j to ʒj entire, or in powder.

CORNUS FLORIDA. Dogwood. *C. circinnata* and *C. sericea* are similar.
Tonic, astringent.
Use. In all cases to which Peruvian Bark is adapted, which it closely resembles, especially intermittents.
Dose. May be given in powder, decoction, or extract of the powder, from ℈j to ʒj. Infusion most employed.

CORYDALIS FORMOSA. Turkey Corn.
Alterative tonic. Used in syphilis, scrofula, etc.
Dose. Of fluid extract 10 to 40 drops.

CORYLUS ROSTRATA. Beaked Hazel.
Anthelmintic. Used the same way as cowhage.

COTULA. Mayweed. Wild chamomile. Herb of *Anthemis Cotula.*
Antispasmodic, vesicant externally. Best given in infusion.

COTYLEDON UMBILICUS. Navel-wort.
Nervous tonic. Used in epilepsy.
Dose. A tablespoonful of the juice daily.

CREASOTUM. Creasote.
Use. Externally applied in rheumatism and neuralgia. Given in some stomachic affections, as dyspepsia, and anorexia, and to allay nausea and vomiting; used externally in porrigo scutulata, and to relieve toothache; also to foul ulcers and cancerous sores.

Dose. From ♏j to ♏ij.

CRETA PRÆPARATA. Prepared Chalk. Take of chalk a convenient quantity; add a little water to it, and rub it into fine powder; throw this into a large vessel nearly full of water, stir briskly, and after a short interval pour the supernatant liquor, while yet turbid, into another vessel. Repeat the process with the chalk remaining in the first vessel, and set the turbid liquor by, that the powder may subside. Lastly, pour off the water, and dry the powder.

Internally antacid; externally absorbent.

Use. In diarrhœa from acidity; externally when sprinkled over burns, after the inflammation has subsided, and a poultice applied, the skinning over the sore is much hastened.

Dose. Gr. x to ʒj or more.

CROCUS. Saffron. (*C. sativa.*)
Stimulant, exhilarating; diaphoretic, emmenagogue.

Use. In hysteria and other nervous affections; chiefly to impart color to officinal tinctures.

Dose. Gr. v to ʒss.

CUBEBÆ. Cubebs. (*Piper Cubebæ.*)
Stimulant, purgative, diuretic.

Use. In gonorrhœa, gleet, leucorrhœa. Also as a grateful stomachic, and carminative in disorders of the digestive organs. Cubebs have been recommended in every stage of gonorrhœa, but they are most safe and effectual in chronic diseases, and where the inflammation is confined to the mucous membrane of the urethra. If not speedily useful, they should be discontinued.

Dose. From gr. x to ℥ss. of the powder, every six hours. The volatile oil is sometimes substituted in the dose of ten or twelve drops, suspended in mucilage, or sugar and water.

CUCURBITA CITRULLUS. Watermelon.

Seeds used in strangury, and as a diuretic. Infusion ℥j or ℥ij to Oj, ad lib.

CUMIN SEED.

Aromatic stimulant.

Dose. 15 grs. to ℥ss.

CUNILLA MARIANA. American Dittany.

Used in warm infusions to promote perspiration, relieve colic, dysmenorrhœa.

CUPRI ACETAS. Acetate of copper.

Tonic, stimulant, escharotic.

Use. In epilepsy, chorea, and other spasmodic affections.

Dose. Gr. ¼ gradually increased to gr. ij.

Incomp. Alkalies, chalk mixture, sulphuric acid.

CUPRI SULPHAS. Sulphate of Copper.

Tonic, emetic, astringent, escharotic, alterative, styptic, antispasmodic.

Use. In epilepsy, hysteria, and intermittent fever; and to produce vomiting in incipient phthisis, in croup, and in poisoning; externally as a stimulant to ulcers and to take down fungus. A weak solution is sometimes used as a collyrium in ophthalmia, and as an injection in gleet.

Dose. As a tonic, gr. ¼ to gr. ij in a pill; gr. ij to gr. x in f℥ij of water vomit.

Incomp. Alkalies, earths and their carbonates; sodæ biboras; salts of lead; acetate of iron; acetate and diacetate of lead; astringent vegetable infusions, decoctions and tinctures.

CUPRUM AMMONIATUM. Ammoniated Copper.
Tonic, antispasmodic.
Dose. ¼ to ½ gr. in pill.

CURCUMA. The root of Turmeric. (*C. Longa.*)
Stimulant, tonic.
Use. In debilitated states of the stomach; intermittent fever; dropsy.
Dose. From ʒss of the powder to ʒij, three tablespoonfuls, three times a day, of an infusion made with ʒiij of the root in Oj of water.

CYCLAMEN EUROPŒUM. Sow-bread.
Drastic cathartic.
Dose. 20–40 grs.

CYDONIUM. Quince Seed.
Mucilaginous. Used in poultice in conjunctivitis.

CYNARA SCOLYMUS. Garden Artichoke.
Diuretic.
Dose. Of tincture ʒj, of extract 5 grs.

CYNOGLOSSUM OFFICINALE. Hound's Tongue.
Used as demulcent and sedative.

CYPRIPEDIUM PUBESCENS. Ladies' Slipper.
Tonic, nervine, antispasmodic. In hysteria, etc.
Dose. Of fluid extract ʒss. to ʒj.

CYTISUS LABURNUM. Laburnum.
Purgative, emetic, poison.

DAJAKOCH. Arrow poison of Borneo.
Acts by poisoning the sympathetic ganglia of the heart. The *Upas antiar* of Java, on the heart itself.

DECOCTA. Decoctions.
See Pharmaceutical preparations, p. 162.

DELPHINIUM. Larkspur. *D. consolida.*
Diuretic, emmenagogue, laxative.
Dose. 10 drops of tincture, ʒj to Oj.

DIANTHUS CARYOPHYLLUS. Clove-pink.

Slightly astringent. Used for coloring and flavoring syrup.

DIGITALINUM. Digitalin.
Dose. $\frac{1}{50}$ or $\frac{1}{60}$ gr.

DIGITALIS. Foxglove. (*D. purpura.*) Stimulant, but afterwards sedative, diuretic, narcotic. In overdoses it occasions vomiting, purging, vertigo, delirium, convulsions, and death.

Use. In inflammatory diseases; phthisis; active hemorrhages, and dropsies, unattended by palsy and unsound viscera. From its influence in lowering the pulse, digitalis has been much employed in palpitations and other affections of the heart, in mania, epilepsy, etc., also as an antispasmodic in pertussis and spasmodic asthma; where there is a laxness of fibre, and pale countenance.

Dose. Gr. j in a pill, united with ammoniacum, soap, calomel, or opium, every six or eight hours till the remedy acts by the kidneys.

DIOSCOREA VILLOSA. Wild Yam.
Antispasmodic; said to be specific in bilious colic.
Dose. Of *dioscorine* 1 to 6 grains.

DIOSPYROS. Persimmon. (*D. Virginiana.*)
Astringent, tonic.
Use. The decoction of the bark, in intermittents, and in the form of a gargle in ulcerated sore throat. The fruit, when green, is excessively astringent, and the juice may be advantageously employed where an astringent effect is desired.

DRACONTIUM. Skunk Cabbage. (*Symplocarpus foetidus.*)
Stimulant, antispasmodic, expectorant, narcotic.
Use. In asthma, chronic catarrh, rheumatism, hysteria, epilepsy, whooping-cough, and dropsy. In large doses it produces nausea and vomiting, with headache, vertigo, and dimness of vision.

Dose. Gr. x to gr. xx of the powdered root three or four times a day. It may also be given in infusion or syrup, in doses of from f ʒj to f ʒiv.

DULCAMARA. Woody Nightshade, Bittersweet. (*Solanum Dulcamara.*)

Diuretic, sudorific, narcotic, alterative.

Use. In chronic rheumatism, humoral asthma, dropsy, lepra, scrofula, and jaundice.

Dose. ℈j to ʒj, in powder; in the form of extract, gr. v to gr. x. An overdose produces vomiting and delirium.

ELATERIUM. Wild Cucumber. (*Momordica elaterium.*)

Violently cathartic, hydragogue, diuretic.

Use. In dropsies.

Dose. Gr. $\frac{1}{10}$th to gr. $\frac{1}{4}$ in a pill, or $\frac{1}{2}$ gr. every hour till it operates; or gr. j dissolved in ʒj alcohol, with 4 drops of nitric acid, of which from 30 to 40 drops may be given in water.

ELEMI. *Canarium commune?*

A resin analogous to turpentines.

EMPLASTRA. Plasters.

See Pharmaceutical preparations, p. 177.

EPIGÆA REPENS. Trailing Arbutus. Ground Laurel.

Used in the same way as Uva Ursi and Buchu.

EPILOBIUM AUGUSTIFOLIUM. Willow-herb.

Astringent tonic.

EQUISETUM HYEMALE. Horsetail. Scouring-rush.

Diuretic, used in infusion.

ERECHTHITES HIERACIFOLIA. Fire-weed.

Used in dysentery.

ERGOTA. Spurred Rye.

Stimulant, acting chiefly on the muscular system of the uterus. Narcotic; a narcotico-acrid poison.

Use. In parturition when the pains languish, and the uterine action becomes torpid, provided the os uteri be fully dilated, and the membranes ruptured. In leucorrhœa and uterine hemorrhage.

Dose. ∋j to ℨss. in cases of parturition; gr. v to gr. x in leucorrhœa, three or four times a day. The most common way of giving Ergot is in decoction, ℨj of it bruised to ℥vj boiling water — boil ten minutes; strain and sweeten, and give one-third every half hour — in parturient cases. Or, of the Tincture made by digesting ℥ss. in ℥vj Rectified Spirit four days, ℨj — of the oil, from twenty to fifty drops.

ERIGERON. Fleabane. (*E. Canadense.*)

Diuretic, tonic, astringent.

Use. In dropsy and diarrhœa. Recommended in gravel and nephritic diseases, as well as dropsy.

Dose. Of the powder, from ℨss. to ℨj. Of the infusion, prepared in the proportion of ℥j of the leaves to Oj boiling water, from f℥ij to f℥iv. Aqueous extract, from gr. v to gr. x every few hours.

ERODIUM CICUTARIUM. Storksbill.

Used in decoction in dropsy.

ERYNGIUM. Button Snakeroot. (*E. aquaticum.*)

Diaphoretic, expectorant, emetic.

Use. As an expectorant in pulmonary and catarrhal affections; its effects resemble those of Seneka Snakeroot.

ERYTHRONIUM. Dog's Tooth Violet. (*E. Americanum.*)

Emetic.

Dose. From gr. xx to gr. xxx of the powdered recent bulb, proves emetic; a smaller dose, expectorant.

ERYTHROXYLON COCA. Coca.

Leaves chewed in S. America. Nervous stimulant — extraordinary stories told of its effects in sustaining strength without food.

Dose. ʒij to ʒiv in infusion.

EUCALYPTUS GLOBOSUS. Australian Gum.

Has been used in intermittents, etc.

EUONYMUS ATROPURPUREUS. Burning Bush. Wahoo.

Tonic, laxative, alterative, diuretic, expectorant.

Dose. As a diuretic in dropsy; a wineglassful of decoction or infusion (ʒj to Oj). Fluid extract ʒj to ʒij.

EUPATORIUM. Thoroughwort. Boneset. (*E. perfoliatum.*)

Tonic, diaphoretic, emetic, aperient, according to dose.

Use. As a diaphoretic in catarrh and rheumatism; in intermittents, and remittents, and inflammatory diseases; as a tonic in dyspepsia and general debility; given cold.

Dose. As a tonic, from Ꝫj to ʒj of the powdered leaves, or f℥j to f℥iv infusion; as a diaphoretic, every two hours the infusion should be given warm, while the patient is covered in bed; as emetic and cathartic, a strong decoction, in doses of Oss. or more.

EUPHORBIA. Spurge. (*E. corollata.*)

The root is emetic and cathartic. In small doses, diaphoretic and expectorant. Inferior to ipecacuanha as to safety, and to antimony as to certainty. Externally vesicant.

Dose. Of the powder from gr. x to gr. xx; as a cathartic, from gr. iij to gr. x.

EXTRACTA. Extracts.

See Pharmaceutical preparations, p. 164.

FEL BOVINUM. Ox Gall (Inspissated).

Tonic, laxative.

Use. In cases of deficient bile.

Dose. Grs. v to x.

FERMENTUM. Yeast.

Externally used in poultices.

FERRI CHLORIDUM. Sesquichloride of Iron.
Internally used in tincture. Styptic in solution.
FERRI CITRAS. Citrate of Iron.
A pleasant chalybeate.
Dose. 5 grs. or more.
FERRI ET AMMONIÆ CITRAS. Citrate of Iron and Ammonia.
A pleasant and soluble chalybeate.
Dose. 5 grs. several times a day. May be given with carbonated alkalies, and in effervescence with citric acid.
FERRI ET AMMONIÆ SULPHAS. Ammonio-ferric Alum.
Tonic and astringent.
Dose. 3 to 15 grs.
FERRI ET AMMONIÆ TARTRAS. Tartrate of Iron and Ammonia.
A mild chalybeate.
Dose. 10 to 30 grs.
FERRI ET POTASSÆ TARTRAS. Potassio-tartrate of Iron.
A slightly laxative chalybeate.
Dose. 10 grs. to ℨss. in solution.
FERRI ET QUINIÆ CITRAS. Citrate of Iron and Quinia.
Combines the virtues of both bases.
Dose. In pill or solution 5 or 6 grs. (equal to 1 gr. of Quinia) 3 or 4 times a day.
FERRI FERROCYANIDUM. Prussian Blue.
Tonic, febrifuge, alterative.
Use. Intermittents, epilepsy, neuralgia.
Dose. 3 to 5 grs.
FERRI IODIDUM. Iodide of Iron.
Tonic, emmenagogue.
Dose. Gr. j to viij.
Use. In all cases of debility, in scrofula, incipient

cancer, amenorrhœa, secondary syphilis, mesenteric obstructions. A bad form of the preparation, which should only be kept in solution.

FERRI LACTAS. Lactate of Iron.
Use. In chlorosis, etc.
Dose. 1 to 2 grs., gradually.

FERRI OXIDUM HYDRATUM. (Hydrated Oxide of Iron. Hydrated Sesquioxide of Iron.) Solution of Tersulphate of Iron Oj, Aquæ Ammon. Water, āā. q. s. Mix the solution with Oiij Water, and add Aq. Ammon. till in excess. Wash the precipitate till nearly tasteless. Mix the precipitate with water to measure a pint and a half.
Use. An antidote for poisoning with arsenic and its salts; acts by combining with arsenious acid, and rendering it insoluble.
Dose. ʒj frequently repeated. This preparation of iron will remove arsenic from its solution in water, by adding 12 grains of it for every grain of the arsenic. Of course, it must be given in large quantities, and proportioned to the quantity of arsenic taken.

FERRI PHOSPHAS. Phosphate of Iron.
Use. A valuable tonic in amenorrhœa, and some forms of dyspepsia; also in intermittents.
Dose. Gr. v to gr. x.

FERRI PYROPHOSPHAS. Pyrophosphate of Iron.
An excellent chalybeate.
Dose. 2 to 5 grs. in pill, water, or syrup.

FERRI REDACTUM. Powder of Iron.
Dose. 3 to 6 grs. in pill or powder.

FERRI SUBCARBONAS.
Tonic, emmenagogue, alterative.
Use. It is advantageously employed in tic douloureux and other forms of neuralgia, dyspepsia, chlorosis, chorea, and lately has been much recommended in cancer. One of our best chalybeates.

Dose. Gr. v to ℨss. united with myrrh, bitter extract, or some aromatic.

Incomp. Acids and acidulous salts.

FERRI SULPHAS. Sulphate of Iron.

Tonic, emmenagogue, astringent, anthelmintic; in large doses emetic.

Use. In diseases of general debility, amenorrhœa, with a weak, languid pulse; diabetes; in clysters against ascarides.

Dose. Gr. j to gr. v, combined with myrrh, ammoniacum, and bitter extracts.

Incomp. The earths, chloride of calcium, chloride of barium, alkalies and their carbonates, biboras sodæ, nitras argenti, acetas plumbi, soaps, tannin.

FERRI SULPHAS EXSICCATA. Dried Sulphate of Iron.

Used for making pills. 3 grs. equal to 5 of crystal.

FERRI SULPHURETUM. Sulphuret of Iron.

Used for preparing sulphuretted hydrogen, by dissolving in dilute sulphuric or muriatic acid.

FERRUM. Iron.

Tonic, deobstruent; anthelmintic; producing fetid eructations, owing to its meeting with acid in the stomach, which oxidizes it, and evolves sulphuretted hydrogen gas.

Use. In general debility, dyspepsia, hysteria, chlorosis, worms, and in passive hemorrhages. It can prove useful only when it is oxidized, which is known by the eructations and black fæces.

Dose. Of the filings gr. v to ℈j, with some aromatic powder; or in the form of electuary with honey; or pills with extract of gentian.

Quevenne's Metallic Iron (Ferrum per Hydrogen) —the most useful form. *Dose.* Gr. ij in pill after each meal.

FERRUM AMMONIATUM. Ammoniated Iron.

An aperient chalybeate.

Use. Amenorrhœa, epilepsy, scrofula, etc.

Dose. 4 to 12 grs. several times a day.

FICUS. Figs.

Nutritious, laxative, demulcent.

FILIX MAS. Male Fern Root. (*Aspidium filix mas.*) Anthelmintic.

Use. In tinea lata, and cucurbitina; but perhaps more is to be attributed to the active purgatives with which it is generally followed.

Dose. ʒij to ʒiij of the solid part of the powdered root taken in the morning, and soon after it a strong cathartic of gamboge or jalap, worked off with green tea. This was Madame Nouffer's celebrated remedy.

FŒNICULUM. Fennel. (*F. vulgare.*)

Carminative, diuretic.

Use. In flatulencies.

Dose. ℈j to ʒj, bruised, to Oj boiling water.

FRASERA. American Columbo. (*F. Walteri.*)

A mild and valuable tonic.

Use. In all cases where a pure tonic is needed.

Dose. Of the powder, from ʒss. to ʒj; of the infusion, made with ℥j of the bruised root to Oj boiling water, ℥j to ℥ij several times a day.

FRAXINUS EXCELSIOR. Ash.

Used in gout and rheumatism. ℥j of leaves in Oss. boiling water 3 times a day.

FUCUS VESICULOSUS. Bladder wrack.

Has been used in scrofula, etc., and for diminishing obesity.

Dose. 20 grs. of powder, or equivalent of extract, 3 times a day.

FUMARIA OFFICINALIS. Fumitory.

A decoction of leaves in hepatic and skin diseases.

GALANGAL. Galanga.

Stimulant, aromatic.

Dose. 15 to 30 grs.

GALBANUM. Galbanum Gum-Resin.

Internally antispasmodic, expectorant; externally resolvent, discutient.

Use. In hysteria, particularly that which attends irregular and deficient menstruation; chlorosis, externally to indolent tumors.

Dose. Gr. x to ʒj in pills, or emulsion.

GALEGA OFFICINALIS. Goat's rue.

Diaphoretic, anthelmintic.

GALIUM APARINE. Cleavers. Goose-grass.

Expressed juice aperient, diuretic, antiscorbutic.

Dose. ℥iij, twice a day.

GALLÆ. Galls.

Powerfully astringent, tonic.

Use. They have been used in diarrhœa, intestinal hemorrhages, and intermittents; but they are principally employed in gargles and injections; and the powder to form an ointment for piles, in the proportion of ʒij to lard ℥ij, and powdered opium ʒj.

Dose. When exhibited internally, gr. x to ℈j, twice or thrice a day.

Incomp. Lime-water, potassæ carbonas, plumbi acetas, et diacetatis, cupri sulphas, argenti nitras, ferri iodidum, ferri sulphas, antimonii potassio-tartras, hydrargyri nitras, hydrargyri bichloridum, infusum cinchonæ, solution of isinglass, solution of opium; all of which precipitate the infusion of galls.

GAMBOGIA. Gamboge.

Drastic cathartic, emetic, hydragogue, anthelmintic.

Use. In visceral obstructions and dropsy. In tapeworm, with carbonate of potassa.

Dose. Gr. ij to gr. x in powder, with calomel, pills, etc.

GAULTHERIA. Partridge Berry. (*G. procumbens.*)
Stimulant, cordial, astringent, emmenagogue.

Use. In diarrhœa, amenorrhœa; but chiefly to flavor other medicines.

Dose. Of the infusion, f℥ij to f℥iv; oil, ♏ij to ♏x.

GELSEMIUM. Yellow Jasmine. (*G. Sempervivens.*)
An excellent febrifuge.

Used in neuralgia, headache, chorea, etc.

Dose. Fluid extract, 3 to 20 drops. Tincture 10–30 drops.

GENTIANA. Gentian root. (*G. lutea.*)
Tonic, stomachic, in large doses aperient; antiseptic.

Use. In dyspepsia, hysteria, jaundice; gout, united with aromatics; chlorosis, with chalybeates; and dropsies, with squill and neutral salts. Externally in putrid ulcers.

Dose. Gr. x to ℈ij. Vide Infusion, etc.

GENTIANA CATESBÆI. Blue Gentian.
Similar to the last.

GERANIUM. Crane's Bill. (*G. maculatum.*)
A powerful astringent.

Use. Diarrhœa, and in the second stage of dysentery after evacuants; cholera infantum; passive hemorrhages. An elegant remedy in cases of infants, or of persons with very delicate stomachs. Locally, to indolent ulcers, an injection in gleet and leucorrhœa, a gargle in relaxation of the uvula and aphthous ulcerations of the throat.

Dose. Of the powder, from gr. xx to gr. xxx; of the decoction, from ℥j to ℥ij. It may be given to children, boiled in milk.

GEUM. Water Avens. (*G. rivale.*)
Tonic, astringent.

Use. In diarrhœa, leucorrhœa, passive hemorrhages, general debility.

Dose. Of the powdered root, from ℈j to ʒj three times a day; of the decoction, made with ℥j of the root to Oj of water, from f℥j to f℥ij; a weak decoction is sometimes made by invalids as a substitute for coffee.

GILLENIA. Indian Physic. American Ipecac. (*G. trifoliata.*)

Emetic, cathartic; in small doses tonic.

Use. As a mild emetic where such medicines are indicated; as a substitute for ipecacuanha.

Dose. Of the powdered root, as an emetic, from gr. xx to gr. xxx, repeated every twenty minutes, till it operates; as alterative and tonic, from gr. v to gr. xv.

GLECHOMA HEDERACEA. Ground-Ivy.

In chronic diseases of lungs and kidneys.

Dose. ʒss. to ʒj.

GLOBULARIA ALYPUM. Wild Senna of Europe.

Cathartic and tonic.

Dose. ℥j in decoction.

GLYCERATES.

See Pharmaceutical preparations, p. 173.

GLYCERINA. Glycerine. Sweet principle of oils.

Demulcent, antiseptic.

Use. Externally, in skin diseases, etc.

GLYCYRRHIZA. Liquorice root. (*G. glabra.*)

Demulcent.

Use. In catarrh; but it is generally combined with other mucilages, and is a pleasant and useful demulcent.

Dose. Of the powder, ʒss. to ℥j.

GNAPHALIUM MARGARITACEUM. Life-everlasting.

Used as tea in pectoral and bowel complaints, and as poultice in bruises, etc.

GOLD. *Aurum.*

The preparations of gold are powerfully alterative, and have been but little studied.

GOSSYPIUM. Cotton. (*G. herbaceum.*)

Used in burns and to blisters, but often acts as an irritant in such cases. A solution of gun-cotton in ether forms collodion. A fluid extract used as emmenagogue and abortive.

Dose. Of tinct. ʒj.

GRANATUM. Pomegranate Bark and Flowers, and bark of the root. (*Punica granatum.*)

Astringent, anthelmintic.

Use. In chronic and colliquative diarrhœas, and the protracted stage of dysentery; for tapeworm; externally, as an injection in leucorrhœa, and gargles in angina.

Dose. In substance, ʒss.; of a decoction f℥ss., every three hours.

Incomp. Sulphate of iron, iodide of iron, nitrate of silver, acetates of lead.

GRINDELIA ROBUSTA.

Used in California for asthma; also as an antidote to effects of poison-oak.

Dose. A wineglassful of syrup from decoction.

GUACO. (*Mikania Guaco.*)

Used as antidote to serpent-bites, etc.

Dose. ʒss. to ʒj of tinct.

GUAIACI RESINA ET LIGNUM. Guaiacum Resin and Wood. (*G. officinale.*)

Stimulant, diaphoretic; in large doses purgative.

Use. In chronic rheumatism, gout, cutaneous diseases, and the sequela of lues venerea.

Dose. To produce its first effects, gr. v to ℈j in pills, or in emulsion made with mucilage or yolk of egg; to purge, gr. xv to ʒj, in the same form.

Incomp. The mineral acids.

GUANO. Bird Manure.

Has been used as cataplasm in chronic inflammation.

GUTTA-PERCHA. *Isonandra gutta.*
Used for utensils, bandages, splints, etc.
A solution in bisulphuret of carbon as an artificial cuticle. (See Liquor Gutta-Percha.) Also as a vehicle for caustics.

HÆMATOXYLON. Logwood. (*H. campechianum.*)
Astringent, tonic.
Use. In the protracted stage of diarrhœa and dysentery, under the form of decoction. (℞. Of the shavings ℥j, water Oij. Boil to Oj, and strain.)
Dose. f℥j to f℥ij every three or four hours.
Incomp. The mineral acids, acetic acid, solution of alum, sulphate of iron and of copper, acetate of lead, antimonii potassio-tartras. Opium, decoction of cinchona flava.

HAMAMELIS VIRGINICA. Witch-hazel.
Astringent, sedative, discutient.
Used in hemorrhages and piles. For the latter, equal parts of this bark, white oak bark, and bark of the apple-tree, in decoction, made up with lard.

HEDEOMA. Pennyroyal. (*H. pulegioides.*)
An aromatic stimulant, diaphoretic, diuretic, emmenagogue.
Dose. Of infusion, ad libitum. Oil from ℳj to ℳx.

HEDERA HELIX. Ivy.
Stimulant and emmenagogue.

HELEBORUS FŒTIDUS. Bear's-foot.
Anthelmintic.
Dose. 5 grs. to ℈j of dried leaves, or in decoction.

HELENIUM AUTUMNALE. False Sun-flower.
Errhine.

HELIANTHEMUM. Frostwort. (*H. Canadense.*)
Tonic and astringent.
Dose. 2 grs. of extract.

HELLEBORUS. Black Hellebore Root. (*H. niger.*)

Cathartic, hydragogue, emmenagogue.

Use. In mania and melancholia, dropsy, and in suppression of the menses in plethoric habits; but it may be questioned whether it is equal to jalap, etc. It is seldom obtained genuine.

Dose. Gr. x to ℈ij purge strongly; to produce its other effects, gr. ij to gr. iij, three times a day. Seldom used in substance.

HELONIAS DIOICA.   False Unicorn.   Stárwort.

In atony of generative organs.

Dose. ʒj of powdered root.

HEMIDESMI RADIX.   Indian Sarsaparilla.

Tonic, diuretic, alterative.

Dose. Wineglassful of infusion (ʒij to Oj).

HEPATICA.   Liverwort.   (*H. Americana.*)

Demulcent, slightly tonic, astringent, diuretic, has no very active virtues.

Use. In chronic coughs, hæmoptysis and hepatic affections. The empirical preparations of this plant owe their efficacy to opium, which they contain in considerable quantities.

HERACLEUM.   Masterwort.   (*H. lanatum.*)

Stimulant, carminative.

Use. In epilepsy, attended with flatulence and gastric disorder.

Dose. ʒij to ʒiij of the powdered root daily, long continued, with a strong infusion of the leaves at bedtime.

HEUCHERA.   Alum Root.   (*H. Americana.*)

Very astringent.

Use. Where astringents are indicated; as a local application to ulcers and cancer; also as a styptic.

HIBISCUS ABELMOSCHUS.   *A. moschatus.*

Used in perfumery. *A. esculentus* is cultivated as *okra*, or *gombo*, for culinary purposes.

HIERACIUM VENOSUM.   Rattlesnake-weed.

Supposed antidote to rattlesnake bite.

Dose. Wineglassful of infusion (℥ij to Oj).

HIRUDO. The leech.

Properties well known.

HORDEUM. Barley.

Demulcent, nutritious.

HUMULUS. Hops. (*H. lupulus.*)

Narcotic, anodyne, diuretic.

Use. In gout and rheumatism; under the form of infusion in the proportion of ℥ss. to Oj of boiling water; but the extract is preferable. The powder, formed into an ointment with lard, is said to ease the pain of open cancer. A pillow stuffed with hops is an old mode of procuring sleep in the wakefulness of delirious fever. Its power has been overrated.

Dose. Gr. iij to ℈j united with ʒss. of cinnamon water, twice or thrice a day; of the infusion, ℥jss.

HURA BRASILIENSIS. *Assacon.*

Emeto-cathartic.

HYDRANGEA ARBORESCENS. Hydrangea.

Proposed as a specific in gravel, or beginning of calculi.

Dose. Fluid extract, ʒj to ʒij.

HYDRARGYRI CHLORIDUM CORROSIVUM. Corrosive Sublimate.

Stimulant, antisyphilitic, alterative.

Use. In venereal complaints with the greatest advantage, when a quick and general action is required; but its effects are not permanent. In lepra, combined with antimonials; and in chronic rheumatism. Dissolved in the proportion of gr. iij to water Oj, as a gargle in venereal sore throats; and a little stronger we have found it useful as a gargle in breaking the abscess in cynanche tonsillaris. It is applied externally to tetters, and for destroying fungus; gr. iv in water Oj is a good wash in scabies. It may be given clysterwise, when the stomach will not bear it. Great caution is necessary in using it externally.

Dose. Gr. $\frac{1}{12}$ to $\frac{1}{4}$, made into a pill. When swallowed as a poison, the best antidote is white of egg. (*Orfila.*)

Incomp. Vide Liquor Bichloridi.

HYDRARGYRI CHLORIDUM MITE. Chloride of Mercury, or Calomel. (A chloride by sublimation.) Calomelas. (In prescribing, it is perhaps safest to use the term Calomelas.)

Antisyphilitic, alterative; in large doses purgative.

Use. In venereal diseases and chronic hepatitis. combined with opium; in scrofula with cicuta; in convulsive affections with opium, camphor, assafœtida, etc.; in dropsies with squill, foxglove, and elaterium; and in rheumatism and lepra with antimonials, guaiacum, and other sudorifics. As a purgative in any case not attended with intestinal inflammation; generally united with purgatives, as gamboge, scammony, jalap, or rhubarb.

Dose. Gr. j to gr. ij, night and morning, in a pill; if it do not purge it gradually excites ptyalism; gr. iij to gr. x purge. Children bear larger doses than adults proportionally. A powder made by triturating it with 10 times its weight of sugar acts as a laxative in doses of 10 grs., and 1 to 2 grs. as a mild alterative in cholera infantum, etc.

Incomp. Nitric and hydrochloric acids, alkalies and their carbonates, lime-water, soaps, sulphurets, iron, lead, copper. The bicarbonates of the alkalies do not decompose it.

HYDRARGYRI CYANIDUM. Cyanide of Mercury.

Dose. $\frac{1}{16}$ gr.

HYDRARGYRI IODIDUM VIRIDE. Iodide of Mercury. Protiodide of Mercury.

Excitant, alterative.

Use. In strumous affections and lepra; as an external application. The iodides of mercury are among our most powerful alteratives, uniting in their effects the properties of both their constituents. They affect the mouth more speedily than other mercurials, and are particularly indicated in scrofula and secondary syphilis, in scrofulous habits. Externally, they are used successfully in ulcers, ill-conditioned sores, swelled joints, where we wish to promote the action of the absorbents; and neuralgic affections.

Dose. Gr. $\frac{1}{2}$ to gr. ij, in pill or dissolved in alcohol.

HYDRARGYRI IODIDUM RUBRUM. Red Iodide of Mercury. (Biniodide of Mercury.)

Dose. Gr. $\frac{1}{16}$ to $\frac{1}{4}$.

HYDRARGYRI OXIDUM RUBRUM. Red Oxide of Mercury. Red precipitate.

Used externally.

HYDRARGYRI OXIDUM NIGRUM. Black Oxide of Mercury.

In scrofula, cutaneous affections, and as an alterative in venereal diseases.

Dose. Gr. $\frac{1}{4}$ to $\frac{1}{2}$.

HYDRARGYRI SULPHAS FLAVA. Yellow Sulphate of Mercury. (Turpeth Mineral.)

Prop. A lemon-yellow powder, almost insoluble in water; entirely dissipated by heat, sulphuric acid being evolved, and metallic globules sublimed.

HYDRARGYRI SULPHURETUM NIGRUM. Sulphuret of Mercury. (Ethiops Mineral.)

Alterative.

Use. In scrofula and cutaneous diseases.

Dose. Gr. v to ℥ss.

HYDRARGYRI SULPHURETUM RUBRUM. Cinnabar.

Sometimes used in fumigation as a sialagogue.

HYDRARGYRUM AMMONIATUM. Ammoniochloride, or White Precipitated Mercury. White precipitate. (A binoxide, combined with bichloride of mercury and ammonia, forming a triple salt.)

Detergent.

Use. As an external application, united with lard. In scabies, and some other cutaneous affections.

HYDRARGYRUM CUM CRETA. Mercury with Chalk. (A protoxide, formed by trituration with carbonate of lime.) Take of Mercury ℥iij, Prepared Chalk ℥v. Rub together till all the globules disappear.

Alterative, antisyphilitic.

Use. In porrigo, and other cutaneous affections; in venereal complaints its operation is so slow and weak as to merit no attention. An alterative in visceral diseases of children, especially in chronic diarrhœa and cholera infantum.

Dose. Gr. v to ʒss., twice a day, in any viscid substance.

Incomp. Acids and acidulous salts.

HYDRARGYRUM CUM MAGNESIA. Mercury with Magnesia. (A protoxide, formed by trituration with carbonate of magnesia.)

In every respect this preparation resembles the former; the employment of the carbonate of magnesia instead of chalk, does not alter the properties nor the virtues of the remedy.

HYDRASTIS CANADENSIS. Yellow Root.

Tonic, astringent.

Use. In infusion, as collyria, to old ulcers, in gonorrhœa, gleet, etc.

Dose. Injection of infusion. (℥ss. Hydrastis, in powder, ℥viij Cold Water) four or five times a day, after urination, in gonorrhœa and gleet. fʒj of fluid extract, in dyspepsia, etc.

HYDROCOTYLE ASIATICA. Thick-leaved Pennywort.
  Diuretic.
  Dose. ℥j to Oj of boiling water, daily.

HYOSCYAMI FOLIA ET SEMINA. Henbane Leaves and Seeds. (*H. niger.*)
  Narcotic, anodyne, antispasmodic, slightly stimulant.
  Use. In epilepsy, hysteria, palpitation, palsy, mania, and scirrhus, as a substitute for opium to procure sleep in nervous habits, pertussis, asthma, catarrh, gout, rheumatism, externally as a cataplasm in cancer and glandular swellings; and to dilate the pupil, or in fine powder sprinkled on cancerous sores, to allay pain.
  Dose. Gr. iij to gr. x of the powder; but generally the extract is preferred.

HYPERICUM PERFORATUM. St. Johnswort.
  Astringent.
  Used in domestic practice in doses of ℥ij of the summits.

HYPOPHOSPHITES. The hypophosphites of lime and soda have been extensively used in tuberculosis, etc. For defects in the osseous system, they seem well adapted.
  Dose. 10–30 grs. 3 times a day.

HYSSOPUS OFFICINALIS. Hyssop.
  Stimulant, aromatic. Used in infusion.

IBERIS AMARA. Bitter Candy-tuft.
  Used in rheumatism, asthma, and dropsy.
  Dose. Of seeds 1–3 grs.

ICHTHYOCOLLA. Isinglass. Sounds of the swimming-bladders of fishes, as the Weak Fish and Cod, but especially the different species of sturgeon.
  Nutritive, demulcent, externally adhesive.

Use. As a diet for the sick and convalescent, and infants troubled with acidity of the primæ viæ. As an article of diet and in cholera infantum, far preferable to vegetable farinaceous substances, as arrowroot, etc. The English court-plaster is made with it.

Incomp. Astringent vegetable infusions, carb. potash, alcohol.

IGNATIA. Ignatia bean.
Similar to Nux Vomica.
Dose. Of Extract ½ gr. to 1 gr.

ILEX OPACA. American Holly.
The bitter principle, *ilisin*, has been proposed as a substitute for quinia.

IMPATIENS FULVA. Touch-me-not.
An ointment made by boiling the plant in lard, used in piles.

IMPERATORIA OSTRUTHIUM. Masterwort.
Stimulant, aromatic.

INFUSA. Infusions. See Pharmaceutical preparations, p. 155.

INDIGO.
Has been used in epilepsy, etc.
Dose. ℈j to ʒj.

INULA. Elecampane. (*I. helenium.*)
Tonic, diuretic, expectorant.
Use. In dyspepsia, paralysis, dropsies, asthma.
Dose. ℈j to ʒj in powder.

IODIDE OF AMMONIUM.
Similar to Iodide of Potass.
Dose. 1–3 grs.

IODIDE OF ANTIMONY.
Alterative.
Dose. ¼ gr. to 1 gr.

IODIDE OF BARIUM.
Alterative.
Dose. ⅛ gr. to 2 grs.

IODIDE OF CALCIUM.
Used in tuberculosis, etc.
Dose. 1–4 grs.

IODIDE OF SILVER.
Substitute for internal use of Nit. Silv.
Dose. 1–2 grs.

IODIDE OF SODIUM.
A substitute for Iod. Potass.

IODIDE OF STARCH.
Vehicle for large doses of Iodine.

IODIDE OF ZINC.
Externally to enlarged tonsils (10–20 grs. to f ℥j water).

IODINIUM. Iodine.
Stimulant, absorbent, emmenagogue, alterative.

Use. In bronchocele and other glandular swellings, not of scirrhous nature, scrofula, dropsy, cutaneous diseases, secondary syphilis, rheumatism, gout, hepatitis; to bring on menstruation in young females in whom it has not occurred; to assist the cicatrization of venereal ulcers.

Dose. From gr. ⅛ to gr. i, in solution with Iod. Potass.

IODOFORM. Teriodide of Formyle.
As ointment, suppository, etc., in painful affections.
Anodyne, in addition to virtues of Iodine.

IODO-HYDRARGYRATE OF POTASSIUM. — (Iod. Potass. grs. iiiss., Biniodide Mercury grs. ivss., Aqua destil. f ℥j. Dissolve first the red Iod. Merc., then the Iod. Potass. This solution contains grs. viij.)

Use. In pulmonary complaints, dyspepsia, amenorrhœa, etc. Increases all the secretions. A most admirable remedy.

Dose. 2 to 10 drops of the solution three times a day in syr. sarsaparilla.

IODO-TANNIN.

Used externally in solution, and internally in form of syrup.

IPECACUANHÆ RADIX. Ipecacuan Root. (*Cephælis Ipecac.*)

Emetic in large doses; sudorific, expectorant, in smaller.

Use. To produce vomiting in the commencement of fevers, phthisis, inflammatory diseases, buboes, swelled testicles, and before the paroxysms of ague; to excite nausea in dysentery, asthma, pertussis, hemorrhages, pneumonia, and combined with opium, to produce diaphoresis in rheumatism, gout and febrile disorders.

Dose. For the first intention, gr. xx, alone, or united with tartar emetic gr. j; for the second, gr. j to gr. iij; and the third, gr. ij to gr. vj, with opium gr. j.

Incomp. Vegetable acids, astringent vegetable infusions.

IRIS FLORENTINA. Florentine Orris.

Peculiar fragrant odor, bitterish, acrid taste.

Oper. Cathartic, emetic, diuretic.

Use. In dropsy; but chiefly used for its fragrance in tooth-powder, to correct an offensive breath; to keep up a discharge from issues in the form of small round balls.

IRIS VERSICOLOR. Blue Flag.

Cathartic, emetic, diuretic.

Use. But seldom employed, owing to the distressing nausea and prostration it occasions.

Dose. Dried root, gr. x to gr. xx.

JALAPA. Jalap. (*Ipomea jalapa.*)

Cathartic; the resinous part gripes violently.

Use. In costiveness, mania, worms, and as a hydragogue in dropsy. It is also a good adjunct to quicken the operation of the chloride of mercury, and other purgatives of slow operation. A drop of essential oil prevents its griping.

Dose. Gr. x to ʒss. in pills or a bolus.

JEFFERSONIA DIPHYLLA. Twin-leaf.

Not unlike seneka in effect.

JUGLANS. Butternut. (*J. cinerea.*)

Cathartic; operating without pain or irritation, resembling rhubarb.

Use. In habitual costiveness; fevers, combined with calomel; hepatic diseases with dandelion.

Dose. Gr. xx to gr. xxx as a purge, gr. v, laxative.

JUNIPERUS. Juniper Fruits and Tops. (*J. Communis.*)

Diuretic, carminative, diaphoretic.

Use. In dropsies; but they cannot be depended on alone, although they are an admirable adjunct to digitalis and squills.

Dose. ℈j to ʒss., triturated with sugar, three or four times a day. The best form of exhibiting the fruit is an infusion made with ℥iij bruised, and boiling water Oj.

JUNIPERUS VIRGINIANA. Red Cedar.

Stimulant, emmenagogue, diuretic, diaphoretic.

Use. In amenorrhœa, chronic rheumatism, dropsy; externally, as an irritant ointment, made by boiling the fresh leaves in twice their weight of lard, and adding a little wax; or the dried leaves may be mixed with six times their weight of resin cerate. Applied to blistered surfaces to keep up a purulent discharge; inferior to the savine.

KALMIA LATIFOLIA. Mountain Laurel.

Has been used in diarrhœa, syphilis, and cutaneous eruptions; but is too dangerous.

KINO. *Pterocarpus marsupium.*

Astringent.

Use. In obstinate chronic diarrhœas; uterine, intestinal, and pulmonary hemorrhages, fluor albus.

Dose. Gr. x to gr. xx in powder; or in solution of the powder ʒj, mucilage of gum f ʒj, cinnamon water f ʒv; two tablespoonfuls occasionally. Vide *Tinct.*

Incomp. The mineral acids, alkalies and their carbonates; acetates of lead, nitrate of silver, tartar emetic, sulphate of iron, bichloride of mercury.

KOOSO. Flowers of *Brayera anthelmintica.*

An Abyssinian vermifuge of great repute.

Dose. ʒss. to an adult, followed by a cathartic.

KRAMERIA. Rhatany Root. (*K. triandra.*)

Astringent, diuretic, detergent.

Use. In dysentery attended with bloody stools; in ulceration of the gums, and as a stomachic in dyspepsia.

Dose. ℈ss. to ʒj in powder.

LACTUCARIUM.

Narcotic, diaphoretic.

Use. In coughs, phthisis pulmonalis, and all painful affections.

Dose. From grs. ij to grs. vj.

LAPPA. Burdock. (*L. minor.*)

Aperient, diaphoretic.

Dose. Of decoction (ʒij bruised root in Oiij water boiled to Oij) Oj during the day.

LAVANDULA. Lavender Flowers. (*L. vera.*)

Stimulant, slightly errhine.

LEONTICE THALICTROIDES. Blue Cohosh.

Diuretic, diaphoretic, anthelmintic, exerts a special influence on the uterus.

Useful in chronic uterine disease.

Dose. Fluid extract, 15 to 40 drops.

LEONURUS CARDIACA. Motherwort.

Infusion or decoction used in amenorrhœa, etc.

LEPTANDRA VIRGINICA. Culver's Physic.

Violent cathartic, emetic.

LIATRIS SPICATA. Button Snakeroot.
Diuretic, cholagogue, laxative.
Dose. Fluid extract, ʒj to ʒij.

LIGUSTICUM LEVISTICUM. Loveage.
Stimulant, aromatic. Test for limestone water, which it turns blue.

LIMONES. Lemons. (*Citrus limonum.*)
Juice refrigerant, antiseptic; bark and oil excitant.
Use. The juice as a beverage, diluted with water and sweetened, is useful in febrile and inflammatory complaints, cooling and quenching thirst: alone or combined with wine, in scorbutis; with camphor mixture, decoction of cinchona, or wine, in putrid sore throats, remittent fevers, diabetes, and lienteria; and with common salt in dysentery and colics.
Dose. fʒij, or more, two or three times a day; diluted ad libitum.

LINIMENTA. Liniments. See Pharmaceutical preparations, p. 174.

LINUM. Flaxseed. (*L. usitatissimum.*)
Demulcent, emollient.
Use. In catarrh, dysentery, strangury, etc., as enema and as poultice.
Dose. Of decoction, ad libitum.

LIQUORES. Solutions. See Pharmaceutical preparations, p. 157.

LIRIODENDRON. Tulip-Tree Bark. (*L. tulipifera.*)
Tonic, diaphoretic, stimulant.
Use. In intermittents, chronic rheumatism, dyspepsia.
Dose. Of the powder, ʒss. to ʒij. Infusion, fʒj to fʒij.

LITHIÆ CARBONAS. Carbonate of Lithia.
Solvent of uric acid.
Dose. 3–6 grs., best in carbonic acid water.

LITHIÆ CITRAS. Citrate of Lithia.
  Dose. 5–10 grs.
LOBELIA. Indian Tobacco. (*L. inflata.*)
  Emetic, purgative, expectorant, antispasmodic.
  Use. In the paroxysms of asthma; in croup, whooping-cough.
  Dose. In powder, gr. iv to gr. xx; infusion, f℥j, tincture ♏xv to ♏xxx.
LONICERA CAPRIFOLIUM. Honeysuckle.
  A syrup of the flowers used in asthma, etc.
  Fruit emetic and cathartic.
LYCOPUS. Bugle Weed. (*L. Virginicus.*)
  Narcotic, tonic, diaphoretic.
  Use. In affections of the lungs, quiets irritation, allays cough, diminishes the pulse.
  Dose. Of the infusion, ad libitum.
LYTHRUM SALICARIA. Purple Willow-herb.
  Demulcent and astringent.
  Dose. Of powdered herb, ʒj. Decoction, (℥j to Oj,) f℥ij.
MAGNESIA. Magnesia. (Obtained from Carbonate of Magnesia, by exposure to a strong heat.) Magnesia Usta.
  Antacid; laxative when it meets with acids in the stomach.
  Use. In heartburn, aphthæ, and other acidities: preferable to chalk when the bowels are costive. Sometimes it is given in dysentery, combined with ipecacuanha and opium, and followed by successive draughts of lemonade.
  Dose. Grs. x to ʒj occasionally, in water or milk.
  Incomp. Acids, metallic salts; hydrochlorate of ammonia.
MAGNESIÆ CARBONAS. (Prepared from Sulphate of Magnesia by Carbonate of Soda.)
  Antacid; laxative when it meets with acid.

Use. The same as that of magnesia; but, owing to the carbonic acid, it sometimes occasions unpleasant distension.

Dose. f℥ss. to ℥ij in water.

MAGNESIÆ SULPHAS. Sulphate of Magnesia. (Obtained from sea-water: magnesian limestone.)

Purgative, diuretic.

Use. In all cases which require purgatives. It operates without griping, and, when united with infusion of roses acidulated, will sit on the stomach when all other things are rejected. The less it is diluted, if a draught of warm water be taken an hour afterwards, the better and more easily it operates. An adjunct to clysters.

Dose. ʒss. to ʒj. In clysters, ʒjss. to ʒiij.

Incomp. The fixed alkalies and their carbonates, lime-water, chloride of barium, nitrate of silver, acetates of lead.

MAGNOLIA. Magnolia. (*M. glauca,* etc.)

A gently stimulating aromatic tonic, and diaphoretic.

Use. In intermittents, chronic rheumatism, and gastric debility.

Dose. Of the powdered bark, ʒss. to ʒj often repeated. The infusion is less efficient.

MALAMBO. Matias or Winter's Bark.

Aromatic, tonic, and febrifuge.

MALVA. Common Mallow. (*Malva sylvestris.*)

Demulcent, similar to Linum.

MANDRAGORA OFFICINALIS. Mandrake.

Poisonous, narcotic. Sometimes used externally to painful tumors.

MANGANESE. The salts of this metal have been advantageously combined with those of iron, in anæmia, etc.

Doses. Of Iodide, 10–30 drops of syrup: Of Phosphate, 1 gr.: Of Lactate, 1–5 grs.

MANGANESII OXIDUM. Oxide of Manganese.

Use. In syphilis, scurvy, itch, and porrigo.

Dose. 3 to 20 grs. in pill, Ointment ℨij to ℨj Lard.

MANGANESII SULPHAS. Sulphate of Manganese.

A cholagogue purgative in dose of ℨj–ℨij; as a tonic, 5–20 grs.

MANNA. Manna. (*Ornus Europœa.*)

Laxative, apt to gripe.

Use. As a purgative for children, who readily take it on account of its sweetness; but more generally it is used as an adjunct to other purgatives.

Dose. ℨss. to ℨij alone, or dissolved in fluid purgatives.

MARANTA. Arrowroot. (*A. arundinacea.*)

When boiled with water or milk, it forms a mild, nutritious article of food, well adapted for infants and convalescents; a tablespoonful into Oj of water.

MARRUBIUM. White Horehound. (*M. vulgare.*)

Tonic, diuretic, laxative, emmenagogue.

Use. In hysteria, chronic catarrh, and pituitous asthma; obstruction of the catamenia; seldom used.

Dose. In powder ℨss. to ℨj; of the expressed juice, fℨss. to fℨjss.; or of this infusion (Marrub. Fol. ℨss., Aquæ Ferv. Oj) a large glassful twice or thrice a day.

MASTICHE. Mastich (resinous tears of *Pistacia*).

Use. Formerly in place of turpentine; now chiefly in ethereal solution, to stop carious teeth.

MATICO. *Piper augustifolium.*

Aromatic, tonic, stimulant, styptic.

Dose. ℨss. to ℨij, three times a day.

MATRICARIA. German Chamomile (flowers of *M. Cham.*).

Similar to chamomile in effects.

MEL. Honey.

Aperient, externally detergent; stimulant.

Use. Seldom used internally as a medicine; but when freely eaten is apt to produce colic; externally as an adjunct to gargles in cynanche tonsillaris; in aphthæ; sometimes applied to foul ulcers.

MEL DESPUMATUM. Prepared Honey. (Take of Clarified Honey Oss., Diluted Alcohol Oj, Prepared Chalk ℨss. Having mixed the honey and diluted alcohol, add the prepared chalk, and allow the mixture to stand for two hours, occasionally stirring it. Then heat it to ebullition, filter, and by means of a water-bath evaporate the clear liquor, so that when cold it may have the specific gravity 1.32.)

MELISSA. Balm. (*M. officinalis.*)
Stomachic, diuretic.
Use. Made into tea; it is used as a diluent in febrile diseases, seldom used in substance.
Dose. Of the powder, grs. x to ƺij.

MEL ROSÆ. Rose Honey. (Rosæ Gallicæ Exsiccat. ℨij, Aquæ Ferv. Oss., Mellis Despum. Oij. Infuse the roses six hours; add the strained liquor to the honey, and evaporate to a proper consistence in a water-bath.)
Astringent, detergent.
Use. Chiefly in gargles, in ulceration, and inflammation of the mouth and fauces (℞. Mellis Rosæ ℨj, Acidi Hydrochlorici ℳ xxx, Aquæ fℨvj; forms a good detergent in aphtha gangrenosa; as a vehicle for other remedies in infantile diseases).

MEL SODÆ BORATIS. Honey of Borax.
Borax 60 grs., Honey ℨj.
Used in aphthæ.

MENISPERMUM CANADENSE. Moonseed.
Substitute for Sarsaparilla.

MENTHA PIPERITA. Peppermint.
Stomachic, carminative.
Use. Vide under Oleum Menthæ Piperitæ.

Dose. Grs. x to ʒj ; scarcely ever in substance.

MENTHA VIRIDIS. Spearmint.

Stomachic, carminative.

Use. Vide under Oleum Menthæ Viridis. An infusion of it is a good diluent in febrile diseases.

Dose. Grs. x to ʒj ; scarcely ever used in substance.

MENYANTHES. Buck Bean. (*M. trifoliata.*)

Tonic, diuretic, purgative; in large doses emetic.

Use. In intermittents, arthritic and chronic rheumatic affections, and in cachectic and herpetic diseases.

Dose. ℈j to ʒj of the dried powdered leaves; f℥j to f℥jss. of the infusion. (Menyanth, fol. sic. ℥ss., Aquæ Oss.)

MESENNA. Bisenna.

An Abyssinian vermifuge for teniæ.

MEZEREUM. Mezereon Bark. (*Daphne mezereum.*)

Stimulant, diaphoretic, in large doses emetic.

Use. In venereal diseases, but its efficacy is doubtful. It is sometimes useful in the sequelæ of syphilis ; in chronic rheumatism, lepra, and scrofulous swellings; and chewing frequently thin slices of the recent root has been found useful in palsy of the tongue; externally, the fresh bark, soaked in vinegar, is useful for keeping open issues.

Dose. Of the powder gr. j, gradually increased to grs. x.

MISTURÆ. Mixtures. See Pharmaceutical preparations, p. 163.

MITCHELLA REPENS. Partridge-berry.

Diuretic, astringent. Similar to Pipsissewa.

MOMORDICA BALSAMINA. Balsam Apple.

Extract said to be useful in dropsy in 6–15 grs.

MONARDA. Horsemint. (*M. punctata.*)

Stimulant, carminative.

Use. In flatulent colic, and sick stomach.

MONESIA. A vegetable extract from S. America.
   Alterative, astringent.
   Dose. 2–10 grs.

MORPHIA. Morphia.
   Narcotic, excitant.
   Use. Chiefly to prepare the more soluble salts. Dissolved in oil, and rubbed upon the skin, it produces narcotic effects.

MORPHIÆ ACETAS. Acetate of Morphia.
   Narcotic.
   Dose. From $\frac{1}{8}$ of grain to gr. $\frac{1}{4}$; endermically, gr. ss. to grs. iij to the skin, where the cuticle has been removed by a blister.

MORPHIÆ MURIAS. Muriate of Morphia.
   As a narcotic, it is preferable to the acetate.
   Dose. Gr. $\frac{1}{6}$.

MORPHIÆ SULPHAS. Sulphate of Morphia.
   Powerfully narcotic and sedative.
   Use. In all cases requiring the use of opium.
   Dose. From gr. $\frac{1}{8}$ to gr. $\frac{1}{3}$.
   ⁎ It is distinguished from sulphate of quinia, which it resembles, by becoming red when treated with concentrated nitric acid.

MOSCHUS. Musk.
   Stimulant, antispasmodic, diaphoretic.
   Use. In spasmodic affections, as hysteria, singultus, pertussis, trismus, and epilepsy. In typhus attended with subsultus tendinum; in cholera it checks the vomiting; and it arrests the progress of gangrene. It raises the pulse and excites the nervous system without heating.
   Dose. Grs. ij to ʒss., every three or four hours.

MUCILAGO ACACIÆ. Mucilage of Gum Arabic. (ʒiv of Pulv. Acacia to Oss. Boiling Water) ʒss. of gum in each fʒ–fʒss. sufficient for a ʒvj or ʒviij mixture.

MUCILAGO TRAGACANTHÆ. Mucilage of Tragacanth. (Gummi Astragali Tragacanthæ Triti ʒij, Aq. Bull. f℥viij. Macerate for twenty-four hours, then triturate till the gum is dissolved, and press through linen cloth.)

Use. In pharmaceutical purposes.

MUCUNA. Cowhage. (*M. pruriens.*)
Vermifuge.
Dose. Mixed with molasses, a teaspoonful to a tablespoonful every morning for three days, followed by a cathartic.

MUSK, ARTIFICIAL.
Antispasmodic.
Dose. 10 grs.

MYRICA CERIFERA. Wax Myrtle. Bay-berry.
Bark tonic and astringent.
Dose. In powder 20–30 grs. Alcoholic extract 5 grains.

MYRISTICA. Nutmegs, Mace, and the Essential Oil. (*M. moschata.*)
Stimulant, stomachic, narcotic in large doses.
Use. To relieve nausea and vomiting, and to check diarrhœa; but chiefly to give flavor to other remedies. Being narcotic, they are hurtful in apoplectic and paralytic habits.
Dose. Of the nutmeg and mace, grs. v to ℈j; of the oil, ♏ij to ♏vj.

MYRRHA. Myrrh. (*Balsamodendron myrrha.*)
Stimulant, expectorant.
Use. In cachectic complaints, humoral asthma, chronic catarrh and phthisis pulmonalis, unattended by hectic or much active inflammation.
Dose. Grs. x to ʒj in powder, united with nitre, camphor, sulphate of potassa, sulphate of zinc, or of iron.

NAPHTHA. Petroleum.
Use. A stimulating antispasmodic and sudorific,

given in disorders of the chest, especially in the West Indies; for the tapeworm in Germany, by mixing one part petroleum with one and a half parts Tinct. Assafœtida, of which 40 drops are given three times a day. Latterly recommended highly in the cure of consumption. Also in cutaneous diseases.

Dose. Mix ℈j naphtha, suspended by a small quantity of boiling alcohol, in ℥iv simple syrup, and give a teaspoonful every fifteen minutes till expectoration is fully established. Mix ℈iij naphtha with ℈xxx lard, and apply in tinea, psoriasis, etc.

British oil is made by mixing the following ingredients: ℞. Olei Terebinth. f℥viij, Olei Lini f℥viij, Olei Succini f℥iv, Olei Juniperi f℥iv, Petrolei Barbadens f℥iij, Petrolei Americana (Seneca Oil) ℥j. Mix.

NECTANDRA. Bebeeru bark. (See Bebeeru.)
Tonic, febrifuge.
Dose. ℈j to ℨj.
NUX VOMICA. Vide Strychnos.
OLEA. Oils. See page 149.
ONION. Garden Onion. (*Allium Cepa.*)
Stimulant, diuretic, expectorant, rubefacient.
Dose. A teaspoonful of juice, with sugar, in non-inflammatory catarrhs of children.
OPIUM. Opium.
Stimulant in small doses, but in larger narcotic, antispasmodic, diaphoretic, sedative, anodyne; externally, its stimulant effects are considerable, but soon followed by its narcotic.

Use. In all painful affections, where the inflammatory diathesis is not very considerable; in diarrhœa and dysentery; intermittents; in typhus, in smaller doses as a cordial, in larger to allay irritation and produce sleep; cholera and pyrosis; in rheumatism when inflammatory fever is not present;

retrocedent gout; and in convulsive and spasmodic diseases. When combined with calomel, in inflammation, after bloodletting, and in syphilis, as well as to arrest the progress of gangrene. It is employed in a watery solution, containing gr. ij in f ℨj of water, as an injection in gonorrhœa and spasmodic stricture, as an adjunct to clysters in diarrhœa; and by friction, united with oil in tetanus and other spasms.

Dose. Gr. ¼ to gr. ss., to produce its stimulant effects; gr. j to grs. ij, is narcotic; but in spasmodic complaints it has been given to a very great extent.

Incomp. Lime-water, alkaline carbonates, bichloride of mercury, nitrate of silver, sulphates of zinc, copper, and iron, infusion of yellow bark, astringent infusions and decoctions, solutions of catechu and of kino; acetates of lead.

⁎ When opium has been taken as a poison, the stomach should be first evacuated by the stomach-pump, worked with infusion of yellow bark, or by emetics containing very little water, and, after the whole of the opium has been evacuated, aromatic stimulants given, and mustard cataplasms applied externally.

As the dose of opium varies much, according to circumstances, and as the quantities vary in pharmaceutical preparations, we have thought it advisable to insert the following table of proportions for reference:

OPIUM. *Dose.* ¼ of a grain to 2 grs. or more.

Acetum Opii (Black drop) contains 1 gr. Opium in ♏vii.

Confectio Opii contains 1 gr. Opium in grs. xxxvi.

Morphia. 1–6th of a gr. equivalent to gr. j Opium.

Morphia Acetas. 1–6th of a grain equivalent to gr. j Opium.

Liquor Morphiæ Acetatis. 1–6th gr. Morphia in ♏v.
Morphiæ Murias. Same as Morphia.
Morphiæ Sulphas. Same as Morphia.
Pilulæ Calomelanos et Opii. 1 gr. Opium to iij grs. Calomel.
Pilulæ Opii. 1 gr. Opium in each pill.
Pilulæ Plumbi Opiatæ. One-half gr. Opium in each.
Pilulæ Saponis Compositæ. 1 gr. Opium in grs. v.
Pulvis Ipecacuanhæ et Opii. 1 gr. Opium in grs. x.
Tinctura Opii (Laudanum). 1 gr. Opium in ♏xix.
Tinctura Opii Camphorata (Paregoric). 1 grain Opium in f ℨss.
Tinctura Opii Acetata. 1 gr. Opium in ♏xx.
Trochisci Glycyrrhiza et Opii. One-tenth grain in each.
Vinum Opii. Same as tincture.

ORIGANUM. Common Marjoram. (*O. vulgare.*)
Tonic, stomachic, emmenagogue?
Use. In debilities of the stomach; scarcely ever used.
Dose. Grs. x to ℈j in powder.

OROBANCHE VIRGINIANA. Beech drops. Cancer-root.
A parasite on the root of the beech. Astringent.

OXALATE OF IRON.
A chalybeate without astringency.
Dose. 2–3 grs. in pill.

OXALIC ACID. Has been used in phthisis, for night-sweats, etc., in dose of ½ gr.
Antidote to poisonous dose, Magnesia or chalk.

OXALIS ACETOSELLA. Wood sorrel.
Refrigerant. Useful in scurvy in infusion.

PÆONIA OFFICINALIS. Peony.
Has been useful in epilepsy.
Dose. Decoction of fresh root ℨij to ℨj boiled from Oj to Oss. daily.

PANAX. Ginseng. Root of *P. quinquefolium.*
Demulcent. The Chinese panacea.

PAPAVER. White Poppy Capsules. (*P. somniferum.*)
Relaxant, anodyne.

Use. Externally as a fomentation (℥iv of the dried heads being bruised and boiled in Oiv of water to Oij), to inflamed or ulcerated parts. The addition of a little distilled vinegar aids the narcotic power of the decoction.

PAREIRA. *Pareira brava.*
Tonic, aperient, diuretic.
Use. In irritable bladder.
Dose. Grs. xxx to ʒj.
Tincture (one part to five of alcohol) fʒj.

PARTHENIUM INTEGRIFOLIUM. Prairie Dock.
Antiperiodic.
Dose. ℥ij in infusion equals ℈j Sulph. Quin.

PAULLINIA. Guarana.
Tonic, nervine.
Dose. 8–10 grs. of Alcoholic Extract.

PEPO. Pumpkin Seed.
Used to expel tape-worm.
Dose. About ℥ij, in emulsion, etc.

PESSARIES, MEDICATED.
Cacao butter, impregnated with medicine for application to the uterus — as Alum 15 grs., Tannin 10 grs., Iodide of Lead 5 grs., Opium 2 grs., Oxide of Zinc 15 grs., Perchloride of Iron 5 grs., etc.

PETROSELINUM. Parsley Root. (*P. sativum.*)
Aperient, diuretic.
Use. In nephritic and dropsical affections; given in infusion. Juice as a substitute for Quinia.

PHOSPHAS SODÆ. Phosphate of Soda.
Purgative.
Use. In all cases where the bowels require to be opened. When dissolved in broth made without salt, the taste of the phosphate is not perceived.

Dose. ʒj to ʒij.

Incomp. Alum, chalk, and all salts with an earthy base.

PHOSPHORUS.

A powerful stimulant, particularly of kidneys and genitals.

Dose. 4-5♏ of a solution of 1 part to 4 of Chloroform.

PHYSOSTIGMA. Calabar bean. (*P. Venenosum.*)

A spinal sedative, gastric irritant; contracts the pupil of the eye.

Use. ⅓ gr. of Alcoholic extract by subcutaneous injection in tetanus.

PHYTOLACCÆ BACCÆ ET RADIX. Poke Berries. Poke Root. (*P. decandra.*)

Emetic, purgative, alterative, and narcotic. A narcotico-acrid poison.

Use. The juice, evaporated to an extract, is employed as an escharotic by cancer doctors. As an alterative in small doses in chronic rheumatism. As an ointment in psora, tinea capitis, and other cutaneous diseases.

Dose. As an emetic, from grs. x to grs. xxx. As an alterative, from gr. j to grs. v.

PILULÆ. Pills. See Pharmaceutical preparations, p. 170.

PIMENTA. Pimenta Berries. (*Myrtus Pimenta.*)

Stimulant, carminative.

Use. Chiefly as a condiment, and as an adjunct to other medicines.

Dose. Gr. v to ℈ij.

PIPER LONGUM. Long Pepper. Similar to Black Pepper.

PIPER NIGRUM. Black Pepper.

Tonic, antiperiodic, stimulant, carminative.

Use. To check nausea in gouty habits; remove

hiccough; and increase excitement in palsy. Steeped in rum it cures ague. A watery infusion of pepper has been found a useful gargle in relaxation of the uvula.

Dose. Gr. x to ℈j, variously combined.

PIX. Pitch.
Stimulant, tonic.
Use. In cutaneous diseases and piles.
Dose. 10 gr. to ʒj in pills; externally as ointment.

PIX BURGUNDICA. Burgundy Pitch.
External rubefacient.

PIX CANADENSIS. Hemlock Pitch.
As a gentle rubefacient, analogous to Burgundy pitch, and employed in the same cases.

PIX LIQUIDA. Tar.
Stimulant, diuretic, sudorific; externally detergent.
Use. Internally in ichthyosis; externally it is applied to foul ulcers, and tinea capitis.

PLANTAGO MAJOR. Plantain.
Refrigerant, diuretic, but feeble.
Used domestically as a dressing for sores.

PLATINUM. The Bichloride of Platinum has been used in syphilis, in the same way as iodine, arsenic, and gold, in ½ gr. doses.

PLUMBI ACETAS. Acetate of Lead.
Astringent in weak solutions, cooling and sedative; in strong (ʒj to water f℥vj) stimulant.
Use. Internally in visceral hemorrhages washed down with water acidulated with distilled vinegar, which seems to prevent its deleterious effects. Externally, in solution in phlegmonous inflammations, burns, bruises, gonorrhœa, etc.
Dose. Gr. ss. to grs. 3, made into a pill with gr. ss. of opium, and crumb of bread. Distilled water must be used for the solution, and a little acetic acid added.
Incomp. Alkalies, earths, acids, alum; borax, soaps,

tartarized iron and antimony; lime-water, hard water, sulphuretted hydrogen.

PLUMBI CARBONAS. Carbonate of Lead.
Astringent, sedative.
Use. Sprinkled on parts affected with local inflammation; in the formation of ointments and plasters.

PLUMBI IODIDUM. Iodide of Lead.
Used in ointment for the effects of its constituents.

PLUMBI NITRAS. Nitrate of Lead.
Externally used as sedative and disinfectant.

PLUMBI OXIDUM. Litharge.
Used in plasters.

PODOPHYLLIN, or resin of podophyllum is much used.
Dose. As laxative $\frac{1}{8}$ to $\frac{1}{4}$ of a grain, as purgative $\frac{1}{4}$ to 1 gr.

PODOPHYLLUM PELTATUM. May Apple. Mandrake.
An active and certain cathartic, producing copious liquid discharges, resembling jalap.
Use. In most inflammatory affections, where brisk purging is indicated; and also in bilious fevers and hepatic congestions; also in dropsical, rheumatic, and scrofulous complaints, in combination with super-tartrate of potassa. A substitute for calomel.
Dose. Of the powdered root, gr. xx. It is also used in the form of an extract.

POLYGALA RUBELLA. Bitter Polygala.
Tonic, laxative, and diaphoretic, according to the dose.
Use. To impart tone to the digestive organs, in the form of infusion.

POLYGONUM PUNCTATUM. Water Pepper.
Stimulant, diuretic, emmenagogue, vesicant.
Dose. Fluid extract 10 to 60 drops.

POPULUS TREMULOIDES. American Poplar.
　Tonic, diuretic, febrifuge.
　Used in intermittents.
　Dose. 4 to 8 grs. of *populin*. The buds of P. balsamifera are often steeped in spirits and applied to bruises, etc. They are balsamic.

POTASSA. Fused Potassa. (Prepared by evaporating the solution of potassa to dryness, in an iron vessel.)
　Powerfully escharotic.
　Use. For forming issues. It has also been used to remove strictures.

POTASSA CUM CALCE. Potassa with Lime. Vienna Caustic. Equal parts of potassa and lime rubbed together — prepared for use by being made into paste with a little alcohol.
　A milder and more manageable caustic than potassa.

POTASSÆ ACETAS. Acetate of Potassa.
　Mildly cathartic, diuretic.
　Use. In febrile diseases, dropsies, icterus, and visceral obstructions.
　Dose. ℈j to ℨj, as diuretic; ℨij to ℨiij open the bowels.
　Incomp. Mineral acids, decoction of tamarinds, bichloride of mercury, nitrate of silver, sulphates of soda and of magnesia, hydrochlorate of ammonia, tartrate of potassa.

POTASSÆ BICARBONAS. Bicarbonate of Potassa.
　Use. The same as that of the carbonate, but it is less acrid.
　Dose. ℈j–℈ij.

POTASSÆ BICHROMAS. Bichromate of Potassa.
　Alterative, emetic.
　Dose. ⅕ gr. daily as alterative, emetic ¾ gr.

POTASSÆ BITARTRAS. Bitartrate of Potash. Cream of Tartar. The tartar of wine purified.

Mildly purgative, refrigerant, diuretic.

Use. In ascites proceeding from visceral obstructions; and to open the bowels in inflammatory habits. Dissolved in water, with a small quantity of white wine, some sugar, and lemon-peel, it forms an excellent beverage in febrile diseases, under the name of Imperial.

Dose. ℈j to ʒj combined with ℈j sodæ biboras, to excite the kidneys; to open the bowels ℨiv to ʒj are required.

Incomp. Alkalies, alkaline earths, mineral acids.

POTASSÆ CARBONAS PURUS. Carbonate of Potassa (pure). Salt of Tartar. Crude carbonate of potassa is *pearlash.*

Diuretic, antacid.

Use. In dropsy, acidities of the primæ viæ, and glandular obstructions.

Dose. Gr. x to ʒss. properly diluted; ℈j dissolved in f℥viij of water, and mixed with f℥iv of lemon-juice, forms an effervescing draught.

Incomp. Mineral acids, borax, hydrochlorate and acetate of ammonia, alum, sulphate of magnesia, chloride of calcium, lime, lime-water, all the metallic salts.

POTASSÆ CHLORAS. Chlorate of Potassa.

Refrigerant, diuretic, etc.

Use. In scurvy, scarlatina, etc., and as a wash in cancrum oris.

Dose. Gr. x to xxx; 1 part to 10 of Glycerin for ulcers, etc.

POTASSÆ CITRAS. Citrate of Potassa.

Refrigerant, diaphoretic.

Dose. 20–25 grs.

POTASSÆ ET SODÆ TARTRAS. Rochelle Salt.

A mild cooling purgative.

Dose. ℥ss. to ℥j.

POTASSÆ NITRAS. Nitrate of Potassa, or Nitre.

Diuretic, refrigerant; in large doses purgative; externally cooling, detergent.

Use. In fevers, dropsies, herpetic eruptions, active hemorrhages, mania. A small piece allowed to dissolve slowly in the mouth often removes incipient cynanche tonsillaris; hence its utility in gargles.

Dose. Gr. x to ʒss. In doses of ʒj it occasions hypercatharsis, bloody stools, and sometimes death.

Incomp. Sulphuric acid, sulphates of soda and magnesia, alum, the metallic sulphates.

POTASSÆ PERMANGANAS. Permanganate of Potassa.

A powerful disinfectant.

Dose. Solution of 10 parts in 90 of water; has been used in petechial fever, diphtheria, etc., in ½ gr. doses.

POTASSÆ SULPHAS. Sulphate of Potassa.

Purgative.

Use. In the visceral obstruction to which children are liable, and as an adjunct to other purgatives.

Dose. ʒss. to ʒvj.

Incomp. Nitric and hydrochloric acids, tartaric acid, chloride of calcium, salts of mercury, nitrate of silver, salts of lead.

POTASSÆ TARTRAS. Tartrate of Potassa.

Purgative.

Use. To open the bowels in febrile diseases, mania, and hypochondriasis; and as an adjunct to senna, and the resinous purgatives in solution, the griping effect of which it corrects.

Dose. ʒj to ʒj in solution.

Incomp. Acids; infusion of tamarinds and other acid fruits: chloride of calcium; lime, magnesia, sulphates of soda, of potassa, and of magnesia; nitrate of silver, acetate of lead, and hydrochlorate of ammonia.

POTASSII BROMIDUM. Bromide of Potassium.

Use. As Iodide of Potassium, but slower in effect. It is also used as a nervous sedative.

Dose. Grs. v to xx.

POTASSII CYANIDUM. Cyanide of Potassium.

Similar to hydrocyanic acid.

Doses. ⅛ gr. in ℥ss. water; f℥ss. syrup of lemons will free the hydrocyanic acid.

POTASSII FERROCYANIDUM. Ferrocyanide of Potassium.

Sedative, anodyne, diaphoretic, astringent; mostly used in the arts.

Use. In dyspnœa, neuralgia, etc.

Dose. 10 to 15 grs. in solution.

POTASSII IODIDUM. Iodide of Potassium.

(Formed by decomposing the iodide of iron by carbonate of potassa.)

The same as that of iodine; but chiefly as an alterative in secondary syphilis, rheumatism, lepra.

Dose. Of the saturated solution from ℳvj to ℳxx; of the dry salt from gr. ij to gr. x. See Tinct. Iodin.

Incomp. Acids, metallic salts, not iodides.

POTASSII SULPHURETUM. Sulphuret of Potassium.

Expectorant, diaphoretic; externally detergent.

Use. It has been given in chronic asthma, but without much benefit; chronic catarrh and rheumatism; arthritic cases; hepatic and other cutaneous diseases; and cancer. Its solution is useful as a wash in scabies and tinea capitis. As a bath, in the proportion of ℥iv to thirty gallons of water; as a lotion in local cutaneous affections, in the strength of ℥j to two quarts of water.

Dose. Gr. ij to gr. x, combined with soap or extract of conium, in pills or mixture, twice or thrice a day; as an ointment, ʒss. of the sulphuret to ℥j of lard.

Incomp. Acids, acidulous salts, metallic and earthy salts.

PRENANTHES SERPENTARIA. Lion's Foot.

Considered a remedy for the bite of the rattlesnake, etc. A decoction of the root is used internally, and the moistened leaves externally.

PRINOS. Black Alder. (*P. verticillatus.*)

Tonic, astringent, alterative.

Use. Intermittents, diarrhœa, gangrene, chronic and cutaneous eruptions; locally in ill-conditioned ulcers.

Dose. Of the powder, from ℨss. to ℨj; of the decoction, made by boiling ℨij of the bark with Oiij of water to Oij, from ℨij to ℨiij; or it may be given in tincture.

PROPYLAMIA. Propylamin.

A volatile alkaloid from herring-pickle, ergot, etc. Has been useful in rheumatism.

Dose. f℥ss. of solution of 24 drops in f℥vi of peppermint-water, every 2–4 hours.

PRUNUM. Prunes. (*P. domestica.*)

Cooling, laxative, nutrient.

Use. In costiveness attended with heat and irritation; an article of diet in fever.

PRUNUS VIRGINIANA. Wild-Cherry Bark.

Tonic and sedative.

Use. In debilitated states of the stomach or general system, attended with irritation and nervous excitability. It allays the action of the heart, and is highly useful in the hectic fever of scrofula and consumption. In dyspepsia and intermittents.

Dose. In powder, from ℨss. to ℨj.

PYRETHRUM. Pellitory. (*Anacyclus pyrethrum.*)

Use. Chewed, it excites a copious flow of saliva, hence it has been found useful in some affections of the head; in strumous swellings of the tonsils; toothache, and palsy of the muscles of the throat. It is also used in infusion as a gargle.

PYRETHRUM PARTHENIUM. Feverfew.
Tonic.
Dose. 3 to 10 grs. three times a day.

QUASSIA. Quassia Wood. (*Simaruba excelsa.*)
Tonic, stomachic.
Use. In intermittents; bilious fever, combined with neutral salts; lienteria and cachexia; in hysteria united with tincture of valerian; and with cretaceous powder and ginger in gout.
Dose. Of the raspings grs. v to ʒss.; but infusion and extract are preferable forms of exhibiting it.

QUERCUS. The Oak.
Astringent, tonic. Not much used internally.
Use. As astringent wash, gargle, or injection.

QUINIÆ MURIAS. Muriate of Quinine. (Dissolve pure quinine in diluted muriatic acid, and evaporate.)
A tonic, better adapted in cases of weak digestive powers than the sulphate; preferred by some to the sulphate in intermittents.
Dose. The same as the sulphate.

QUINIÆ SULPHAS. Bisulphate of Quiniæ. (Prepared from yellow cinchona.)
Tonic.
Use. In intermittents and all periodic diseases, as a tonic; also as a febrifuge in bilious remittents, and whenever tonics are indicated; may be used with great advantage endermically where the stomach is irritable.
Dose. Grs. i to grs. x in any simple bitter infusion.
Incomp. Alkalies and their carbonates, lime-water, salts of baryta, lime, nitrate of silver, and salts of lead.

QUINIÆ VALERIANAS. Valerianate of Quinia.
Use. In neuralgia and hemicrania.
Dose. 1 to 2 grs.

RANUNCULUS. Crowfoot. (*R. bulbosus.*)
  Acrid irritant; similar to cantharides.

RENNET. GASTRIC JUICE. PEPSINE.
  These forms of gastric juice have been found useful in dyspepsia from debility of stomach. Rennet wine made by digesting the lower part of calf's stomach in good sherry is a common form.
  Dose. A teaspoonful in a wineglassful of water.

RESINA. Yellow Resin. (The residue, after the distillation of oil of turpentine.)
  Stimulant.
  Use. In the composition of plasters and ointments.

RHAMNI. Buckthorn. (Berries and juice of *R. catharticus.*)
  Purgative.
  Use. In syrup, added to hydragogue or diuretic mixtures.
  Dose. ℈j of recent berries, ʒj of dried, ʒj of juice.

RHEUM. Rhubarb Root. (*R. palmatum.*)
  Purgative, stomachic, astringent.
  Dose. In costiveness, from laxity of bowels, particularly of children, and diarrhœa. It is a useful adjunct to neutral salts and calomel, rendering their operation more easy. Externally, the powder is sprinkled over ulcers, to assist their granulation and healing.
  Dose. Grs. x to ʒss. of the powder to open the bowels; grs. vj to grs. x, to act as a stomachic.

RHIGOLENE. A variety of petroleum naphtha, used for congelation with the Atomizer.

RHUS GLABRUM. Sumach. (Fruit of *R. glab.*)
  Astringent and refrigerant.
  Used as a gargle.

ROSA. Rose Petals.
  Slightly laxative.

ROSMARINUS. Rosemary. (*R. officinalis.*)

Tonic, stimulant, emmenagogue, resolvent.

Use. In nervous headaches, and in chlorosis, under the form of infusion; but it is now scarcely ever used, unless as an adjunct, to give odor to sternutatory powders.

Dose. Of the powder grs. x to ℨss.

ROTTLERA. Kameela. (*R. tinctoria.*)
Purgative, anthelmintic for tænia.
Dose. ℨj–ℨiij.

RUBIA. Root of Madder. (*R. tinctorum.*)
Emmenagogue, astringent.

Use. In chlorosis, and difficult or scanty menstruation; in the atrophia infantum; but its efficacy is very doubtful.

Dose. Grs. xv to ℈j, united with sulphate of potassa, three or four times a day.

RUBUS TRIVIALIS VILLOSUS. Dewberry Root, Blackberry Root.
Astringent and tonic.

Use. In diarrhœa from debility, cholera infantum, chronic dysentery. In all cases where astringents are indicated.

Dose. Of the decoction (ℨj, Ojss. water; boiled to Oj), from fℨj to fℨij three or four times a day. Of the powdered root, grs. xx to grs. xxx.

RUMEX. Dock-root. (*R. Britannica* and *obtusifolius.*)
Mild astringent, tonic, alterative.

Use. In scrofula and syphilis, externally in skin diseases.

Dose. fℨij of decoction (ℨj dried root to Oj).

RUTA. The leaves of Rue. (*R. graveolens.*)
Tonic, stimulant, antispasmodic, emmenagogue.

Use. In hysteria and flatulent colic; but chiefly in the form of strong infusion in clysters in the convulsions of children.

Dose. Grs. xv to ℈ij.

SABADILLA. Sabadilla Seeds. (*Veratrum sabadilla.*)
Cathartic, excitant, anthelmintic.

Use. Seldom internally; used in the form of powder to destroy pediculi. (Recommended by Turnbull in painful rheumatic and neuralgic affections.)

Dose. Gr. ⅙ of the extract, grs. ij to grs. vj of the powder. Tincture used externally.

SABBATIA. American Centaury. (*S. angularis.*)
Tonic.

Use. In autumnal intermittents and remittents.

Dose. Infusion of ʒj to Oj of water, fʒij every two hours.

SABINA. Savine Leaves. (*Juniperus sabina.*)
Stimulant, diaphoretic, emmenagogue, anthelmintic, escharotic.

Use. In amenorrhœa, with a languid pulse, but they require to be cautiously administered; in worms, rheumatism, and gout. Externally, the powder is applied to old ulcers, carious bones, etc.; and the infusion, as a lotion, to gangrene, scabies, and tinea capitis.

Dose. Grs. v to grs. x of the powder.

SACCHARUM. Sugar.

SACCHARUM LACTIS. Sugar of Milk.

SAGAPENUM.
Antispasmodic, emmenagogue, inferior to assafœtida.

Dose. Grs. x to ʒss. in pills.

SAGO. Sago. (*Sagus Rumphii.* A modification of starch, containing traces of chloride of sodium.)

SALEP. Prepared bulbs of Orchis.
Nutritive, like sago, etc.

SALICINA. Salicine.
Tonic.

Use. In intermittents, and in all cases where tonics are indicated. Its effects are analogous to those of quinine, but not much used.

Dose. Grs. iv to grs. vj every three hours in intermittents. In other cases, gr. j to grs. iij three or four times a day.

SALIX CORTEX. Willow Bark. (*S. alba.*)

Tonic, astringent.

Use. In intermittents and remittents; debilities of the intestinal canal; convalescence; and in hectic and phthisis.

Dose. ℈j to ʒj of the powder; or f℥jss. of the decoction, made with ℥ij of the bark, in Oij water, boiled down to Oj.

Incomp. Solution of isinglass, alkaline carbonates, lime-water, sulphate of iron.

SALVIA. Sage. (*S. officinalis.*)

Tonic, astringent, aromatic.

Use. Mostly as gargle, in infusion, with honey and alum.

SAMBUCUS. (*S. Canadensis.*) Common Elder Flowers, Berries, and Bark.

Flowers diaphoretic, discutient; berries aperient, sudorific; bark purgative, hydragogue.

Use. The flowers in fomentations, to yield their flavor to water in distillation, and to form a cooling ointment; the berries, or their expressed juice, in febrile diseases, rheumatism, arthritic cases, and the exanthemata; the bark in dropsy and hemorrhoids.

Dose. Of the juice of the berries f℥j to f℥ij; of the bark grs. v to ʒss., three times a day.

SANGUINARIA. Blood Root. (*S. Canadensis.*)

An acrid emetic; stimulant, narcotic, diaphoretic, alterative.

Use. It is principally used in chronic catarrh, bronchial affections, and pertussis. Combined with antimony or ipecacuanha, it is a useful expectorant.

Dose. As emetic, from grs. x to grs. xx; as alterative, gr. j to grs. iv. Of the tincture, x to xxx drops. This is the best form of administration.

SANTALUM. Red Saunders Wood. (*Pterocarpus santalum.*)

Used for coloring tinctures.

Oil of yellow Saunders wood (*S. myrtifolium*) in doses of 20–40♏ is said to be specific in gonorrhœa.

SANTONIN. The active anthelmintic principle in European wormseed.

Used in the form of lozenges with sugar.

Dose. Three or four grs. twice a day, or less to a child.

SAPO. Soap.

Laxative, antacid, antilithic.

Use. In dyspepsia, constipation, lithiasis, etc.

Dose. 5 grs. to ℨss., in pill.

Incomp. Acids, earths, earthy and metallic salts.

SAPONARIA OFFICINALIS. Soapwort.

Alterative, like sarsaparilla.

Dose. ℨss. inspissated juice daily.

SARRACENIA. Fly-trap.

The *S. purpurea* has been used in small-pox, to modify or shorten its course; ℨss. of root to Oj water in wineglassful dose every three hours.

SARSAPARILLA. *Smilax officinalis.*

Diuretic, demulcent.

Use. In the sequelæ of syphilis, when, after a mercurial course, nocturnal pains, enlargement of the joints, and cutaneous ulcerations remain; in scrofula; elephantiasis, or cutaneous affections resembling it; chronic rheumatism; and whenever an alterative is indicated.

Dose. From ℈j to ℨj of the powder, or made into an electuary, three times a day. See Decoction, Syrup, and Ext.

SASSAFRAS. Sassafras Wood and Root. (*Laurus Sassafras.*)

Stimulant, sudorific, diuretic.

Use. In cutaneous diseases; chronic rheumatism; and as an adjunct to the decoction of guaiac, etc.

SCAMMONIUM. Scammony. (*Convolvulus scammonia.*)

Drastic purgative, hydragogue.

Use. In obstinate costiveness, worms, dropsy, in combination with some other cathartic, as aloes, rhubarb, calomel, etc.

Dose. Grs. iij to grs. xv triturated with sugar or with almonds. Made into an emulsion with milk, the taste cannot be distinguished; but pure virgin scammony is exceedingly scarce.

SCILLA. The bulb of the squill. (*S. maritima.*)

Emetic in large doses; purgative; in small doses expectorant and diuretic. It owes its properties to a peculiar principle, which has been named scillitin.

Use. In pulmonary complaints, after the inflammatory action is reduced; humoral asthma; pertussis; in dropsy; and more useful if combined with a mercurial.

Dose. Gr. j to grs. v of the dried root, powdered, and united with nitre or ipecacuanha; or in pills to produce diuresis, united with the blue pill.

Incomp. Gelatine, lime-water, alkaline carbonates, acetates of lead, nitrate of silver.

SCOPARIUS. Broom Tops. (*Cytisus scoparius.*)

Diuretic.

Use. In dropsies.

Dose. ℈j to ℨj of the powder. 10–15 grs. of seeds.

SCROPHULARIA NODOSA. Figwort, pilewort, Celandine.

An old English remedy for piles and scrofulous tumors, in ointment or fomentation.

SCUTELLARIA LATERIFOLIA. Scullcap.

A tonic nervine.

Used in tic douloureux, etc.

Dose. Fluid extract, ʒss. to ʒj.

SELINUM PALUSTRE. Marsh Parsley.

Used in epilepsy, etc.

Dose. 20–30 grs. three times a day.

SENECIO AURENS. Life-root.

Diuretic, diaphoretic, tonic.

Useful in gravel and uterine complaints; promotes menstruation.

Dose. Fluid extract, ʒss. to ʒj.

SENEGA. Senega Root. (*Polygala senega.*)

Stimulant, expectorant, diaphoretic, diuretic.

Use. In peripneumonia, after the inflammatory action is reduced; humoral asthma, chronic rheumatism; dropsy; croup? The extract of it, with carbonate of ammonia, has been found useful in lethargy.

Dose. Grs. x to ℈j of the powder. Madeira wine, if it can be ordered, covers the taste of the powder.

SENNÆ FOLIA. Senna Leaves. (*Cassia acutifolia.*)

Cathartic, hydragogue. (It is apt to gripe.)

Use. In costiveness and dropsy; should always be given with aromatic and saline substances.

Dose. Of the powder, ℈j to ʒj rubbed with crystals of bitartrate of potassa, and united with ginger to prevent griping; but the best form is that of infusion.

SERPENTARIÆ RADIX. Snake Root. (*Aristolochia serpentaria.*)

Stimulant, diaphoretic, diuretic.

Use. In typhoid fevers, and diseases of debility; to assist cinchona in the cure of intermittents; in the exanthemata, and dyspepsia; and externally as a gargle in cynanche maligna.

Dose. Of the powder, grs. x to ʒss.; or of the following infusion f℥ss. every four hours: ℞. Rad. contusi Serpentariæ ʒiv, Aquæ ferv. f℥xij. Macerate in a covered vessel for two hours.

SESAMUM. Benne. (*S. indicum.*)
Laxative, demulcent, nutritious.

Use. As a drink in cholera infantum, diarrhœa, dysentery, catarrh, and affections of the urinary passages.

Dose. One or two green leaves in a tumbler of cool water will render it sufficiently viscid.

SEVUM. Suet.
Used in ointments, etc.

SIMARUBA. The Bark and Wood of *Simaruba officinalis*.
Tonic.

Use. In dysentery, chronic diarrhœa, lienteria, and dyspepsia.

Dose. ℨss. to ℨj of the powder; but the infusion is a better form of exhibiting this remedy.

SINAPIS. Mustard Seed. (*S. alba* and *nigra.*)
Stimulant, diuretic, emetic, rubefacient, laxative.

Use. In dyspepsia; a torpid state of the bowels, and chlorosis. The seeds are swallowed entire or only slightly crushed; a strong infusion of the flour is used to produce vomiting in apoplexy and paralysis; externally the flour is applied as a cataplasm to the legs and the soles of the feet in typhus and comatose affections.

Dose. ℨj to ℥ss. or f℥ij of the following infusion: ℞. Sinapis pulveris, Armoraciæ rad., sing. ℥ij, Aq. ferventis Oij. Infuse in a covered vessel for twelve hours; then strain and add Spir. Menthæ Piper. f℥ij.

SODÆ ACETAS. Acetate of Soda.
Purgative, refrigerant, diuretic.

Use. In cases requiring a mild purgative. Chiefly used for making acetic acid.

Dose. From ℈j to ℨiv in any bland fluid.

Incomp. Carbonate of lime, sulphuric, nitric, and hydrochloric acids.

SODÆ BICARBONAS. Bicarbonate of Soda.

Use. The same as that of the carbonate.

Dose. Grs. x to ʒss.

SODÆ BORAS. Borate of Soda. Borax.

Diuretic, emmenagogue.

Use. In nephritic and calculous complaints, depending on an excess of uric acid. As a detergent in aphthous affections of the mouth in children, rubbed up in sugar in the proportion of 1 to 7, or rubbed with honey.

Dose. From grs. xx to grs. xl; or combined with cream of tartar.

SODÆ CARBONAS. Carbonate of Soda.

Antacid.

Use. In dyspepsia, and acidities of the stomach; united with bitters; in uric acid and gravel, in whooping-cough, bronchocele, and in scrofulous affections.

Dose. Grs. x to ʒss. twice or thrice a day.

Incomp. Lime, acids, unless as an effervescing draught, hydrochlorate of ammonia, earthy and metallic salts.

SODÆ CARBONAS EXSICCATA. Dried Carbonate of Soda. (The carbonate made to undergo the watery fusion; and, when dry, reduced to powder.)

Antacid, lithontriptic.

Use. In acidity of the stomach; but chiefly in calculus in the kidneys, and other affections of the urinary organs.

Dose. Grs. v to grs. xv. made into pills, with some aromatic powder and soap.

SODÆ HYPOSULPHIS. Hyposulphite of Soda.

Destructive to microscopic fungi, and arrests fermentation.

Dose. 10–20 grs. 3 times a day, in syrup or water.

SODÆ PHOSPHAS. Phosphate of Soda.

A mild purgative.

Dose. ʒj to ʒij in gruel or weak broth.

In small doses (3 to 10 grs.) phosphate of soda acts beneficially in hepatic disorders of children.

SODÆ SULPHAS. Sulphate of Soda, or Glauber's Salts. (From the salt which remains after the distillation of hydrochloric acid, the superabundant acid being saturated with carbonate of soda.)

Purgative; in small doses diuretic.

Use. In costiveness; in bilious colics, largely diluted.

Dose. Of the effloresced salt in powder, ʒiij to ʒvj; of the crystallized salt in solution, ʒvj to ʒxij; its nauseous taste may be corrected by lemon-juice or cream of tartar.

Incomp. Carbonas potassæ, chlorides of calcium and barium, salts of lead, of silver.

SODÆ SULPHIS. Sulphite of Soda.

Antizymotic, same as the hyposulphite.

Dose. ʒj 3 times a day.

SODÆ VALERIANAS. Valerianate of Soda.

Nerve stimulant.

Dose. 1–5 grs.

SODII CHLORIDUM. Chloride of Sodium. Muriate of Soda, or Sea Salt.

Tonic, purgative, anthelmintic, externally stimulant.

Use. In some cases of dyspepsia and worms; in sea scurvy, and purpura; in large doses to check vomiting of blood; as an ingredient in clysters; a fomentation to bruises; and, added to water, to form a stimulant bath.

Dose. Grs. x to ʒss. In clysters ʒiv to ʒj.

SOLIDAGO. Golden Rod. (*S. odora.*)

Aromatic, stimulant, carminative, diaphoretic.

Use. To relieve pain arising from flatulence; to allay nausea. In warm infusion.

SPIGELIA. Indian Pink Root. (*S. marilandica.*)
Anthelmintic.

Use. For the expulsion of lumbrici; in the remitting fever of infancy. Its use should be preceded by an emetic, and followed by a warm purgative.

Dose. Grs. x to ʒss. of the powdered root, every night and morning, till the worms are expelled; or an infusion combined with senna.

SPIRÆA. Hardhack. (*S. tomentosa.*)
Tonic, astringent.

Use. In cholera infantum, diarrhœa, and all cases where a tonic combined with an astringent effect is needed.

Dose. Of the extract, from grs. v to grs. xv; from f℥j to f℥ij of the decoction.

SPIRITUS. Spirits or Essences. See p. 168.

SPIRITUS VINI GALLICI. Brandy.

SPONGIA. Sponge.

Use. External. For absorbing the acrid discharge from ulcers; suppressing hemorrhages, when the bleeding mouth of the vessel is compressed with it; to form tents for dilating wounds, in which case the sponge is immersed in melted wax, and cooled before being used; for making burnt sponge.

SPONGIÆ USTÆ PULVIS. Burnt Sponge. (The sponge is cut into pieces, burnt to a friable coal in a covered vessel, and rubbed to a powder.)

Tonic, deobstruent, antacid.

Use. In bronchocele, scrofulous complaints, and herpetic eruptions.

Dose. ʒj to ʒiij, made into an electuary, with honey and powdered cinnamon.

STANNI PULVIS. Powder of Tin.
Anthelmintic.

Dose. ℥ss. in molasses, for several mornings, followed by a cathartic.

STATICE. Marsh Rosemary. (*S. caroliniani.*)
Astringent, antiseptic.
Use. In gargles, in aphthous and malignant sore throat; and internally in chronic dysentery.

STILLINGIA. Queen's Root. (*S. sylvatica.*)
Emetic, cathartic, alterative.
Use. In secondary syphilis, scrofula, etc.
Dose. Powder, 15 to 30 grs., tincture (℥ij to Oj) f℥j. 20 to 40 drops of fluid extract.

STRAMONIUM. The Leaves and Seeds of Thorn Apple. (*Datura stramonium.*)
Use. The same as the extract.
Dose. Seeds gr. j, powdered leaves grs. ij.

STRYCHNIA. Strychnia. An alkali prepared from the Strychnos Nux Vomica.
Use. As a tonic in pyrosis, passive diarrhœa, and leucorrhœa; in cases of partial paralysis not depending on organic disease, especially when caused by carbonate of lead.
Dose. From gr. $\frac{1}{16}$th to gr. $\frac{1}{12}$th, in acid solution.

STRYCHNIÆ SULPHAS. Sulphate of Strychnia.
More soluble than strychnia.
Dose. $\frac{1}{24}$th to $\frac{1}{12}$th gr. internally, $\frac{1}{8}$ to $\frac{1}{3}$ externally.

STRYCHNOS IGNATIA. Ignatia Bean. (*Ignatia amara.*)
Tonic, and nervous stimulant; virtue depends on strychnia.
Use. In dyspepsia of all grades.
Dose. 5 to 10 drops of fluid extract.

STRYCHNOS NUX VOMICA.
Tonic, stimulant; when taken in large doses, it produces tetanic symptoms.
Use. In dyspepsia, gout, rheumatism; and especially in paralysis of the lower extremities.
Dose. From grs. iij to grs. xij.

STYRAX. Storax. (*S. officinale.*)

Stimulant, expectorant.

Use. Seldom used alone, but as an adjunct, chiefly on account of its fragrance and aromatic properties.

Dose. Grs. x to ℨss.

SUCCINUM. Amber.

Use. To afford its essential oil and acid.

SULPHUR. Sublimed Sulphur. Flowers of Sulphur.

Stimulant, laxative, diaphoretic, transpiring through the cutaneous exhalants.

Use. As a laxative in chronic rheumatism, atonic gout, rachitis, asthma, and some pulmonary affections; in hemorrhoidal affections it is the only laxative that should be employed, united with magnesia or bitartrate of potassa. A specific in itch, and several cutaneous diseases, when either internally or externally exhibited.

Dose. ℨss. to ℨij taken night and morning.

SULPHUR PRÆCIPITATUM. Precipitated Sulphur, Lac Sulphuris.

Laxative and alterative; emmenagogue.

Use. In cutaneous affections, and as a laxative in constipation and hemorrhoids.

Dose. ℨj in the form of an electuary, two or three times a day, or combined with magnesia or cream of tartar.

SULPHURIS IODIDUM. Iodide of Sulphur. (℞. Iodine ℨiv, Sulphur ℨj. Rub together in a glass mortar till thoroughly mixed. Put the mixture into a matrass, close the orifice loosely, and apply a gentle heat, so as to darken the mass without melting it. When the color has become uniformly dark throughout, increase the heat so as to melt the iodide; then incline the matrass in different directions; and, lastly, allow it to cool, break it, and put the iodide into bottles, which are to be kept well stopped.)

A powerful alterative, especially in lupus, acne, and psoriasis.

Use. In cutaneous affections, secondary syphilis, rheumatism, etc. The ointment of iodide of sulphur should be made at first by mixing grains x of the iodide with ʒj lard: the strength may be gradually increased, as the skin can bear it, until it contains ℈ss. to the ʒj lard or spermaceti ointment.

SULPHOCARBOLATES, of soda, zinc, magnesia, etc. are expected to afford a convenient means of obtaining the effect of carbolic acid in diseases of parasitic origin, cholera, and zymotic diseases generally.

SUMBUL RADIX. Sumbul root.
Nerve stimulant.
Dose. Of resinoid extract 1–2 grs. three times a day. Of fluid extract, 15♏–fʒj.

SUPPOSITORIA. Suppositories. See p. 178.

SYMPHYTUM OFFICINALE. Comfrey.
Demulcent, astringent.
Dose. ʒij to ʒiv of fluid extract.

TABACUM. The leaves of Tobacco. (*Nicotiana Tabacum.*)
Narcotic, sedative, diuretic, emetic, cathartic, errhine; a violent poison, whether externally applied or taken into the stomach.

Use. In ileus and incarcerated hernia, in the form of clysters of the infusion, or the smoke in dropsy and dysuria; chewing it relieves the pain of toothache; and as an errhine, it forms the basis of all the snuffs in common use. The infusion has been used as a lotion in scabies, tinea capitis, and other eruptions, but is apt to induce sickness.

Dose. See Infusum Tabaci. For clysters, ʒj is infused in Oj of boiling water.

TAMARINDUS. The Pulp of Tamarind. (*T. indica.*)
Laxative, refrigerant.

Use. In dysentery and fevers, particularly those attended with an increased secretion of bile, and putrid symptoms. Tamarind whey, made by boiling ℨij of the fruit with Ojss. milk, and straining, is an excellent diluent in fevers.

Dose. ʒss. to ʒij, often added to senna and to manna.

Incomp. Carbonates, and acetates of potassa and soda; the resinous cathartics; infusum sennæ.

TANACETUM. Leaves of Tansy. (*T. vulgare.*)

Tonic, anthelmintic.

Use. In gout; hysteria, connected with suppression of the menses; in worms, seldom used.

Dose. ʒss. to ʒj. It is drunk as a tea by gouty people.

TAPIOCA. Tapioca (fecula of root *Janipha Manihot*).

A nutritious diet.

TARAXACUM. The Root of Dandelion. (*Leontodon taraxacum.*)

Aperient, diuretic, resolvent.

Use. In chronic inflammation, and incipient scirrhus of the liver; chronic derangements of the stomach; dropsy, pulmonary tubercles, and jaundice.

Dose. f ℨij of the following decoction three or four times a day: ℞. The full-grown roots, sliced, ℨiv, Water Oij. Boil gently to a pint, strain, and add bitartrate of potassa ʒiij.

Incomp. Infusion of galls, nitrate of silver, bichloride of mercury, acetates of lead, sulphate of iron.

TEREBINTHINA. Turpentine.

Stimulant, diuretic, anthelmintic, laxative, externally rubefacient.

Dose. ƺj to ʒj or ℨss. to ℨj as anthelmintic.

TESTÆ. Oyster shells (burnt).

Antacid, absorbent.

Use. Chiefly in the acidities of infancy, and during dentition.

Dose. Grs. x. to ʒij.

TINCTURÆ. Tinctures. See p. 158.

TORMENTILLA. Tormentil Root. (*Potentilla tormentilla.*)

Astringent.

Use. In the same cases as other astringents; but as it does not increase the heat of the body, tormentil is preferred in phthisical diarrhœa.

Dose. Grs. x to ʒj of the powder; or fℨij of the following decoction: ℞. Pulv. Tormentillæ ℨj, Aq. Puræ Oj, decoque ad fℨxij et cola.

TOXICODENDRON. Poison-oak. (*Rhus toxicodendron.*)

Stimulant and narcotic; an acrid narcotic poison.

Use. In paralytic affections and herpetic eruptions; but in the former its efficacy is doubtful; also in dropsy and phthisis.

Dose. Grs. ss. to grs. iv, twice or thrice a day.

TRAGACANTHA. Tragacanth. (*Astragalus verus.*)

Demulcent.

Use. Small quantities held in the mouth, and swallowed very slowly, sheathe the fauces and allay tickling cough; but it is chiefly used for pharmaceutical purposes, to suspend heavy, insoluble powders, and to impart consistency to troches.

Dose. Grs. x to ʒj.

Incomp. Cupri Sulphas, plumbi acetas, and sulphas ferri, precipitate its mucilage.

TRIFOLIUM PRATENSE. Red Clover.

Recommended in cancerous ulcers.

TRILLIUM PENDULUM. Bethroot.

Astringent, tonic, antiseptic.

Dose. ʒj of powdered root, ʒj to ʒiij fluid ext.

TRIOSTEUM PERFOLIATUM. Fever Root.

Cathartic, emetic, and diuretic.

Use. In the commencement of fevers.

Dose. ℈j to ʒss. of the powder acts as a cathartic; of the extract, grs. x to ℈j. It may be given with advantage combined with calomel.

TROCHISCI. Troches. See p. 167.

TUSSILAGO. Coltsfoot. (*T. farfara.*)

Demulcent, expectorant.

Use. In cough, phthisis, other pulmonary complaints, and cutaneous diseases.

Dose. ʒss. to ʒj in milk. It is more generally given in decoction, made with a handful of the leaves boiled in two pints of water to one pint; strained and sweetened with syrup; the dose, a teacupful occasionally.

ULMI CORTEX. Elm Bark.

Demulcent, feebly tonic.

Used in decoction.

ULMUS. The inner bark of elm. (*U. fulva.*)

Tonic, alterative, diuretic, demulcent, nutritious.

Use. In lepra and other cutaneous affections; diarrhœa, dysentery, diseases of the urinary organs. Externally as an emollient.

Dose. Of decoction ℥iv to ℥vj.

ULMUS FULVA. Slippery-elm Bark.

Demulcent, nutritious.

Used in infusion, or as poultice, etc.

UNGUENTA. Ointments. See p. 175.

UVA PASSA. Raisins.

Laxative.

UVA URSI. Leaves of Bear's Whortleberry. Red-berried Trailing Whortleberry. (*Arctostaphylos uva ursi.*)

Tonic.

Use. In chronic diarrhœa and dysentery; leucorrhœa, and diabetes. It has been celebrated in calculous and nephritic complaints; but it appears to act in the same manner as other astringents, by merely allaying the pain and irritability of the bladder. In Phthisis?

Dose. Of the powder, grs. xv to ℥ss.

Incomp. Salts of iron, tartar emetic, nitrate of silver, salts of lead, infusion of yellow cinchona bark.

VALERIANA.   Valerian Root.   (*V. officinalis.*)

Antispasmodic, tonic, emmenagogue.

Use. Hysteria, epilepsy, hemicrania, chlorosis.

Dose. Of the powder ℈j to ʒj three or four times a day, increasing it as far as the stomach can bear it.

Incomp. Salts of iron.

VALERIANATE OF IRON.

Used in hysterical chlorosis.

Dose. 1 gr. several times a day.

VANILLA.

Aromatic, nerve stimulant.

Dose. Infusion ℥ss. to Oj boiling water, in tablespoonful doses.

VERATRIA.   Veratria.   An alkaloid prepared from Sabadilla.

A powerful topical excitant.

Use. Externally applied as an ointment in neuralgia, and in gouty and rheumatic paralysis.

Dose. Not more than one-twelfth of a grain.

VERATRUM ALBUM.   White Hellebore Root.

Violently emetic; purgative, even when applied externally to an issue; errhine; externally stimulant.

Use. It is never given internally, unless in maniacal cases, in which it is not more useful than other strong purges; and even its use to promote a discharge from the nose in apoplexy and lethargy requires great caution. For its external use, see Ointment.

Dose. As an errhine, grs. iij or grs. iv, snuffed at bedtime.

VERATRUM VIRIDE. American Hellebore.

Slightly acrid, sedative, emetic, diaphoretic.

Use. It is an arterial sedative, in fevers, pneumonia, etc.

Dose. Of the tincture ♏iij to ♏v, repeated every hour or so, and watched till its effects appear on the pulse. The fluid extract is, perhaps, the preferable form.

VERBASCUM THAPSUS. Mullein.

Demulcent, emollient, anodyne. An infusion used in mild catarrhs.

VERBENA OFFICINALIS. Vervain.

Formerly of repute in scrofula.

VIBURNUM PRUNIFOLIUM. Black Haw.

Nervine. Used to prevent miscarriage.

Dose. ʒj–ʒij of infusion or decoction of bark.

VINA MEDICATA. Medicated Wines, see p. 161.

VINUM XERICUM. Spanish White Wine, or Sherry.

When good, and of a proper age, wine, in small quantities, is tonic, antispasmodic, and nutritive; when new, flatulent and purgative; sooner intoxicating, and, instead of strengthening, produces debility.

Use. In the low and sinking stages of typhus fever, the judicious exhibition of it fills the pulse, and restores its firmness, without increasing delirium; but it is hurtful if given when the skin is very hot and dry. It is useful also in tetanus, chorea, and some other convulsive affections; and in most cases in which tonics are indicated. In convalescence from all severe diseases it is a remedy on which much dependence used to be placed; much less used at present. Hock is the best wine for dyspeptics.

Dose. fʒij to Oij in twenty-four hours, according to the nature of the disease and the previous habits of the patient.

VIOLA. Violet. (*V. pedata.*)
 Mucilaginous. Slightly laxative.

WINTERA AROMATICA. Winter's Bark.
 Carminative, tonic.
 Use. As an adjunct to stomachic infusions, in dyspepsia, and scorbutus.

XANTHORRHIZA. Yellow Root. (*X. apiifolia.*)
 Tonic.
 Use. In all cases where a pure tonic is indicated. Its properties are analogous to those of columbo and quassia.

XANTHOXYLUM. Prickly Ash. (*X. fraxineum.*)
 Stimulant, diaphoretic, resembling mezereon and guaiac.
 Use. In chronic rheumatism, and as a topical remedy for toothache.
 Dose. Of the powder from grs. x to ℨss.; of the infusion, from f℥j to f℥iij, three or four times in twenty-four hours; or of the decoction, made by boiling ℥j of the bark in Oij of water for fifteen minutes, f℥iv to f℥viij every three or four hours.

ZINCI ACETAS.
 Use. As an astringent collyrium, and as an injection in gonorrhœa.
 Gr. j to ij to f℥j water.

ZINCI CARBONAS PRÆCIPITATUS. Precipitated Carbonate of Zinc.
 Use. Same as Calamine.

ZINCI CHLORIDUM. Chloride of Zinc.
 Use. As an alterative and antispasmodic in scrofula, epilepsy, etc. As an escharotic in scirrhous tumors, etc.
 Dose. 4 to 8 drops of ethereal tincture (℥ss. to f℥iij).

ZINCI IODIDUM. Iodide of Zinc.
 A solution of 10 to 30 grs. to f℥j of water has been applied with advantage to enlarged tonsils.

ZINCI OXIDUM. Oxide of Zinc.

Tonic, antispasmodic, externally detergent, exsiccative.

Use. In epilepsy, chorea, and other spasmodic affections. For its external use, see Ung. Zinci.

Dose. Gr. j to vj twice a day.

ZINCI SULPHAS. Sulphate of Zinc.

Emetic, tonic, antispasmodic, externally astringent.

Use. As it operates very quickly, it is used, combined with infusion of ipecacuanha, to empty the stomach in the commencement of the cold stage of the intermittent paroxysm; and in other cases where immediate vomiting is required. As a tonic, it is useful in phthisis, dyspepsia, and nervous affections. Externally in collyria, in ophthalmia, after the inflammatory action has subsided; in injections, in gonorrhœa; and as a lotion in external inflammations, and to stop inordinate discharge.

Dose. Grs. x to ℨss., to produce vomiting; as a tonic, gr. j to grs. ij, twice or thrice a day.

Incomp. Alkalies, earths, sesquicarb. ammonia, hydrosulphurets, lime-water, astringent vegetable infusions, milk.

ZINCI VALERIANAS. Valerianate of Zinc.

Antispasmodic.

Dose. 1–2 grs., several times a day.

ZINGIBER. Ginger Root. (*Z. officinale.*)

Carminative, stimulant, sialagogue.

Use. In gout, flatulent colic, dyspepsia, and tympanitis; as an adjunct to griping purgatives; less heating than pepper.

Dose. Grs. x to ∋j; an overdose is apt to induce spasmodic stricture.

VII.

PHARMACEUTICAL ARRANGEMENT OF THE MATERIA MEDICA.

I. — Inorganic Products.

Mineral Acids — Tonics and Astringents.

Acidum Carbonicum (see Aq. Medicata).
"   Muriaticum, HCl,+water. Dose ♏ 3 to 5.
"   "   dilutum. 1 pt. to 3 of water. Dose ♏ 15 to 40.
"   Nitricum. $HO,NO_5 + 3HO$. Dose ♏ 1 to 4.
"   "   dilutum. 1 to 6 pts. water. Dose ♏ 15 to 40.
"   Nitro-muriaticum. 1 pt. Nit., 2 Mur. Acid. Dose ♏ 3 to 5.
"   Sulphuricum. HO,SO_3. Dose ♏ 1 to 2.
"   "   dilutum. 1 pt. to 13 water. Dose ♏ 15 to 40.
"   "   Aromaticum — Alcoholic and Aromatic. Dose ♏ 15 to 30.
"   Phosphoricum (glacial), HO,PO_5. Solid.
"   "   dilutum. 1 pt. to 10 water. Dose ♏ 15 to 40.

The Alkalies and their Salts.

Group 1.

Potash. From lye of wood ashes.
Potassæ Carbonas Impurus. (Pearlash.)
Salæratus. $2KO,3CO_2$.
Potassæ Carbonas. KO,CO_2.
Liquor Potassæ Carbonas. ʒxij to fʒxij water. Antilithic antacid. Dose ♏ 10 to fʒj.
Potassæ Bicarbonas. $KO,2CO_2,HO$.
Liquor Potassæ. Boiling Carb. with Hydrate of Lime. Antacid. Dose ♏ 5 to fʒss.
Potassa. KO,HO. Caustic Potash. Escharotic.
" cum Calce. Milder escharotic.
Potassæ Acetas. $KO,\bar{A}c$. Diuretic, Grs. 10 to ʒij.
" Citras. $KO,\bar{C}i$. Refrigerant, diaphoretic. Dose ℈j to ʒss.
" Chloras. KO,ClO_5. Refrigerant, diuretic. Dose grs. 10 to ʒss.

Group 2.

Sodii Chloridum. NaC. (Common Salt.)
Sodæ Sulphas. NaO,SO_3+10HO. Cathartic. Dose ʒss. to ʒj.
" Carbonas. NaO,CO_2+10HO. (Sal. Soda.)
" Carbonas Exsiccatus. NaO,CO_2. Antacid. Dose grs. 5 to 15.
" Bicarbonas. $NaO,2CO_2+HO$. Antacid. Dose ℈j to ʒj.
" Phosphas. $2NaO,HO,PO_5+24HO$. Cathartic, diuretic. Dose ʒij to ʒj.
Liquor Sodæ Chlorinatæ. Labarraque's Disinfectant.
Sodæ Acetas. $NaO,Ac+6HO$. Used in preparing Acetic Acid.
" Valerianas. NaO,Va. For preparing other valerianates.

Group 3.

Crude Argols or Tartar.   From wine casks.
Potassæ Bitartras. { $KO,HO,^2\overline{T}$ (cream of tartar). Purified by recrystallization. Dose ʒss. to ʒj.
Sodæ et Pot. Tartras. $KO,NaO\overline{T}+8HO$. Cathartic. Dose ʒij to ʒj.
Potassæ Tartras. $2KO,\overline{T}$.   Dose ʒj to ʒj.

Group 4.

Potassæ Nitras. KO,NO_5. Sedative, diuretic. Dose grs. v to ʒj.
Sal prunelle — fused saltpetre.
Potassæ Sulphas. KO,SO_3. Cathartic. Dose ʒj to ʒij.
Sodæ Boras. $NaO,2BO_3+10HO$. Used in gargles, etc.

Group 5.

Ammoniæ Murias. NH_3,HCl. Stimulant, expectorant. Dose grs. 5 to 20.
Liquor Ammoniæ — aqueous sol. of ammonia.
Spiritus   "   · alcoholic   "   " { ♏10 to 30 largely dil.
    "   "   aromaticus   "   with aromatics. Dose ♏20 to fʒj.
Ammoniæ Carbonas. NH_4O,CO_2. Stimulant, antacid. Dose grs. 5.
Liquor Ammon. Acetatis. (Spts. of Mindererus.) Diaphoretic. Dose fʒj to fʒss.

Preparations of Earths.

1. *Lime.*

Creta Preparata. CaO,CO_2. Antacid. Dose grs. 10 to ʒj.

Calcis Carb. Præcipitata — same as above, but more elegant.

Liquor Calcis — Lime-water. Antacid. Dose f℥ss. to f℥ij, in milk.

Calcii Chloridum. CaCl.

Liquor Calcii Chloridii, 1 pt. CaCl in 2.5. Deobstruent. Dose ♏30 to f℥j.

Calx Chlorinata. $CaO,ClO+CaCl$. Chloride of Lime.

Calcis Phosphas. $3CaO,PO_5$. Antiscrofulitic. Dose. grs. 10 to ʒss.

Syrupus Ferri Phosphatis Comp. Syrup of the phosphates. Dose, a teaspoonful.

2. *Magnesia.*

Magnesiæ Sulphas. MgO,SO_3+7HO. Cathartic. Dose. ℥ss. to ℥j.

" Carbonas. $4MgO,CO_2,HO,MgO,2HO$. Antacid. Dose ℈j to ʒj.

" Bicarbonas. (Soluble Magnesia.)

Magnesia. MgO. By calcining the Carb. Cathartic. Dose ʒj.

Liquor Magnes. Citratis. Dose ʒj of salt in f℥xij bottle.

3. *Baryta.*

Liquor Barii Chloridi. ʒj to f℥viij water. Deobstruent. Dose gtt. 5.

4. *Alumina.*

Alumen. $KO,SO_3,+Al_2O_3,3SO_3+24HO$. Astringent, etc.

" Exsiccatum. (Burnt Alum.) Used externally.

INORGANIC PRODUCTS. 143

Non-Metallic Elements.

1. *Iodine.*

Iodinium. I. Alterative.
Potassii Iodidum. KI. Alterative. Dose grs. 2 to 5.
Tinct. Iodinii. ʒss. to f ʒj Alcohol. Used externally.
" " Comp. I, grs. 15. KI. ʒss. to f ʒj. Dose m 15 to 30.
Liquor " I, grs. 22½. KI gr. 45 to f ʒj. Lugol's Solution. Dose m 10 to 20.

2. *Bromine.*

Brominum. Br. Obtained from *bittern* at Salt Works.
Potassii Bromidum. Alterative. Dose grs. 5 to 10.
Liquor Ferri Bromidi. Solution in excess. Alterative. Dose m 5 to 10.

3. *Sulphur.*

Sulphur. S. sublimed. Alterative, laxative. Dose grs. 10 to ʒij.
" præcipitatum. Alterative, laxative. Dose ʒij to ʒiij.
Sulphuris Iodidum. IS_2. In ointment.

Metallic Elements.

1. *Iron.* (Ferrum.)

Ferri Pulvis. Fe. (Quevenne's Iron by Hydrogen.) Dose gr. j to iij.
" Sulphas. $FeO, SO_3 + 7HO$. Hæmetic, astringent. Dose grs. 3 to 5.
" Subcarbonas. $Fe_2O_3 2HO + FeO, CO_2$. Dose grs. 5 to ℈j.

Ferri Phosphas. In Amenorrhœa, etc. Dose grs. 5 to 10.

Tinct. Ferri Chloridi, 32 grs. Fe_2Cl_3 in f℥j alcohol. Astringent. Dose ♏10 to 30.

Ferrum Ammoniatum. Deobstruent. Dose grs. 4 to 10.

Liquor Ferri per Sulphatis. $Fe_2O_3, 3SO_3 + Aq$.

Ferri Oxidum Hydratum. $Fe_2O_3, 3HO$. By adding Ammonia to the above.

" Citras. Fe_2O_3, Ci. Dose grs. 3 to 5.
" et Quiniæ Citras. Dose grs. 2 to 5.
" Lactas. $FeO, L, 3HO$. In Chlorosis. Dose grs. j to iij.
" et Potassæ Tartras. $Fe_2O_3, KO, \overline{T}$. Dose grs. 10 to 20.
" Ferrocyanuretum. $3Cfy, 4Fe$. Sedative, tonic. Dose grs. 5 to 15.

Liquor Ferri Nitratis. $Fe_2O_3, 3NO_5 + Aq$. Astringent. Dose ♏5 to 15.

Ferri Iodidum. FeI. Decomposes. Dose grs. j to ij.

Liquor Ferri Iodidum. FeI. Grs. vij to ℥j Syrup. Dose ♏20 to 40.

Ferri Bromidum. FeBr. Tonic, alterative. Dose grs. 2 to 5.

" Valerianas. $Fe_2O_3, 3Va$. In Hysteria, etc. Dose gr. j.

2. *Manganese.*

Manganesiæ Sulphas. $MnO, SO_3, + 4HO$. Tonic, cath. Dose grs. 5 to ℨij.

" Carbonas. $2MnO, CO_2 + HO$. Dose grs. 5.

Syrupus Manganesii Iodidi. ℨj Mn I to f℥j. Dose ♏10.

3. *Copper.*

Cupri Sulphas. $CuO, SO_3 + 5HO$. Tonic, astringent, etc. Dose gr. ¼ to 5.

Cuprum Ammoniatum. CuO,SO_3+2NH_3HO. Antispasm. Dose gr. ⅕.
Cupri Subacetas. $2\bar{C}uO,\bar{A}c+6HO$. Escharotic.

4. *Zinc.*

Calamina Preparata. (Carbonate.) ZnO,CO_2.
Zinci Sulphas. ZnO,SO_3+7HO. Tonic, ½ gr. to ij; Emetic 10 grs.
" Carbonas Præcipitatus. Used in Cerate.
" Oxidum. ZnO. Tonic, astringent, desiccant.
" Acetas. $ZnO,\bar{A}c$. Astringent; used in Collyria and Injection.
" Chloridum. $ZnCl$. Escharotic, antiseptic.
" Cyanuretum. $ZnCy$. In Epilepsy, Chorea, etc. Dose ¼ to j.
" Valerianas. $ZnO,\bar{A}a$. In nervous affections. Dose grs. j to ij.

5. *Lead.*

Plumbi Oxidum Semivitreum. PbO. Litharge.
" Acetas. $PbO,\bar{A}c,3HO$. Sedative, astringent. Dose grs ½ to iij.
Liquor Plumbi Subacetatis. (Goulard's Extract.)
Plumbi Carbonas. PbO,CO_2. Used externally.
" Nitras. PbO,NO_5. Disinfectant.
" Iodidum. PbI. In resolvent ointment.

6. *Silver.*

Argenti Nitras. AgO,NO_5. (Crystals.) Alterative. Dose gr. ¼ to j.
" Nitras fusus (sticks). Lunar Caustic.
" Oxidum. AgO; a substitute for the Nitrate. Dose gr. ½ to ij.

7. *Bismuth.*

Bismuthi Subnitras. BiO_3,NO_5. Tonic, antispasm.

8. *Antimony.*

Antimonii Sulphuretum. SbS_3. Horse medicine.
" Sulphuretum Præcipitatum. SbO_3+5Sb,S_3+16HO. Alterative. Dose gr. j to iij.
" et Potassæ Tartras. $SbO_3,KOT,+3HO$. Emetic, grs. ij. Diaphoretic and expectorant, gr. $\frac{1}{6}$ to $\frac{1}{4}$. Sedative, $\frac{1}{6}$ to j gr.
Vinum Antimonii. Grs. ij to f ℥j white wine. Dose gtt. x to f ʒj.
Pulvis Antimonialis. (James's powder.) Alterative, diaphoretic. Dose gr. iij to x.

9. *Arsenic.*

Acidum Arseniosum. AsO_3. (White Arsenic.)
Liquor Potassæ Arsenitis. AsO_3,KO,CO_2, 64 grs. each to Oj—grs. iv AsO_3 to f ʒj. (Fowler's Solution.) Dose ♏3 to 15.
Arsenici Iodidum. AsI_3.
Liquor Hydrargyri et Arsenici Iodidum. AsI_3+HgI, each, 70 grs. to Oj. (Donovan's Solution.) Dose ♏5 to 20.

10. *Mercury.*

Hydrargyri Chloridum Corrosivum. $HgCl_2$. Alterative. Dose gr. $\frac{1}{16}$ to $\frac{1}{4}$.
" Chloridum Mite. $HgCl$. Cathartic and alterative. Dose $\frac{1}{12}$ to 20 grs.
" Sulphas Flavus. $3HgO,2SO_3$. Emetic. Dose 3 grs.
" Iodidum Rubrum. HgI_2. Alterative. Dose $\frac{1}{16}$ to $\frac{1}{4}$ gr.

ORGANIC PRODUCTS.   147

Hydrargyri Iodidum. HgI. Viride. Alterative. Dose $\frac{1}{4}$ to 1 gr.
" Sulphuretum Rubrum. HgS. Alterative, fumigations.
" Sulphuretum Nigrum. Hg_2S. Mild alterative. Dose grs. 5 to ʒj.
" Oxidum Rubrum. HgO_2. Stimulant, external.
" Oxidum Nigrum. HgO. Alterative. Dose $\frac{1}{4}$ to 3 grs.
" Cyanuretum. HgCy. Alterative. Dose $\frac{1}{16}$ to $\frac{1}{8}$ gr.
Hydrargyrum Ammoniatum. Hg_2Cl,NH_2. External.
" cum Creta. $3Hg+5,CaO,CO_2$. Antacid alterative. Dose $\frac{1}{2}$ to 3 grs.

II. Organic Products.

Lignin and its Derivatives.

Lignin or Cellulose. $C_{24}H_{20}O_{20}$. (*Gossypium*, cotton.)
Collodium. Ethereal Solution of prepared cotton—Artificial cuticle.
Carbo Ligni (Carbo animalis similar). Charcoal. Dose 1 or 2 teaspoonfuls.
Acidum Aceticum. (Dilutum in 1 pt. to 7 of water.)
Spiritus Pyroxylicus. $C_2H_4O_2$. (Wood Naphtha.) Sedative. Dose 10 to 40 drops.
Creasotum. $C_{14}H_8O_2$. Internally to check nausea. Dose gtt. j.

Farinaceous, Mucilaginous, and Saccharine Medicines.

1. *Fecula.*—Amylum (starch), Canna (tous le mois), Maranta (arrowroot), Sago, Florida arrowroot, Tapioca.

2. *Gums.* — Acacia, Mezquite gum, Salep, Tragacantha.
3. *Sugars.* — Saccharum, Saccharum candium (rock candy), Lactin (sugar of milk), Treacle (molasses), Mel (honey), Manna, Extractum glycyrrhizæ (liquorice).

Protein and Similar Principles.

Fel Bovinum. (Inspissated ox-gall.) Laxative. Dose grs. 5 to 10.

Ichthyocolla (Gelatin) — as dietetic and in plaster.

Alcohol and Ethers.

Alcohol. C_4H_5O,HO. The standard has sp. gravity .835. Of this, Brandy has 55 per cent., Irish Whiskey 52, Rum has 53, Gin 51, strong Port Wine 25, weak Port 19, Currant Wine 20, Madeira 24, Sherry 19, Claret 12 to 17, Hock 12, Champagne 12, Cider 5 to 9, Ale 6 to 8, Porter 4 to 6 per cent.

Æther. C_4H_5O. Best Anæsthetic 1 pt. Ether, 2 of Chloroform.

Spiritus Ætheris Compositus. (Hoffman's Anodyne.) Dose gtt. 20 to f℥j.

Spiritus Ætheris Nitrici. Refrigerant, diaphoretic. Dose gtt. 10 to f℥ij.

Chloroformum. C_2HCl_3. Anæsthetic, anodyne, etc. Dose gtt. 20 to 60.

Fixed Oils and Fats.

Glycerin. $C_6H_7O_5 + HO$ (sweet principle). Lubricant, miscible with water, etc.

List. — Adeps (lard), Oleum Adipis and Stearin (from lard), Sevum (mutton suet), Ol. Amygdalæ, Ol.

ORGANIC PRODUCTS. 149

Macidis (from fruit of myristica moschata), Ol. Cacao, Ol. Olivæ, Ol. Papaveris, Ol. Sesami (benne oil), Oil Lini (flaxseed), Ol. Bubulum, Ol. Morrhuæ (cod-liver), Ol. cetacei, Ol. Ricini (castor oil), Ol. Tiglii (croton), Ol. Palmæ.

Volatile or Essential Oils.

Carbo-Hydrogen essential oils are the oils of Turpentine, Savine, Juniper, Cardamoms, Lemon, Cedrat, Neroli, Bergamot, Orange, Cubebs, Copaibæ, Pepper, Ginger, Amber, Cloves, and Valerian.

Oxygenated oils (most soluble), are oils of Anise, Absinthium, bitter Almonds, Asarum, Achillea, Buchu, Cajeput, Canella, Carraway, Catnip, Cascarilla, Cloves (heavy), Chenopodium, Carrot seed, Cassia, Cinnamon, Chamomile, Coriander, Cumin, Dill, Erigeron, Filix mas, Fennel seed, Gaultheria, Hedeoma, Hops, Lavender, Marrubium, Matico, Matricaria, Melissa, Mint (pepper and spear), Monarda, Nutmeg, Origanum, Pimenta, Pulegium, Rose (attar), Rosemary, Rue, Salvia, Sambucus, Sassafras, Serpentaria, Tanacetum, and Valerian (heavy).

Sulphuretted oils are oils of Mustard, Horseradish, Garlic or Onion, and Assafœtida.

Camphors have a close relation to essential oils; many of which deposit them.

Olea. Oils.

Oleum Amygdalæ Dulcis. Dose f℥j to f℥j.
    " Anthemidis. 5–15 drops.
    " Bergamii. Perfume.
    " Bubulum. Neat's-foot oil. Dose as cod-liver oil.
    " Cajeputi. Stimulant. 1–5 drops.
    " Camphoræ. 1–2 drops. Rubefacient anodyne.

Oleum Cinnamomi. 1-2 drops.
" Limonis. Flavoring.
" Lini. Flaxseed oil.
" Morrhuæ. Cod-liver oil. Dose a tablespoonful.
" Myristicæ. 1-2 drops.
" Oliva. Olive oil.
" Ricini. Castor oil. Dose f ℨj. Give in mint or cinnamon water, or hot coffee.
" Rosæ. Perfume.
" Succini. Oil of Amber.
" Terebinthinæ. Dose 5=30 drops or more.
" Theobromæ. Cacao butter. Used for suppositories, etc.
" Thymi (Origani). Local application.
" Tiglii. Croton oil. Dose 1-2 drops. Counter-irritant.

Distilled Oils.

Oleum Anisi. 5-15 drops.
" Cari. 1-10 drops.
" Caryophylli. 2-6 drops.
" Chenopodii. Anthelmintic dose for child 4-8 drops.
" Copaibæ. 10-15 drops.
" Coriandri. 5-10 drops.
" Cubeba. 10-12 drops.
" Erigontis Canadensis. 5-10 drops.
" Fœniculi. 5-15 drops.
" Gaultheriæ, 5-10 drops.
" Hedeomæ. 2-10 drops.
" Juniperi. 5-15 drops.
" Lavandula. 1-5 drops.
" Menthæ Piperitæ. 1-3 drops.
" " Viridis. 2-5 drops.
" Monardæ. 2-3 drops. Rubefacient.
" Origani. Rubefacient.

Oleum Pimenta. 3–6 drops.
" Rosmarini. 3–6 drops.
" Rutæ. 2–5 drops.
" Sabinæ. 2–5 drops.
" Sassafras. 2–10 drops.
" Succini Rectificatum. 5–15 drops.
" Tabaci. Rarely used.
" Valerianæ. 3–5 drops.

Oleoresins.

Oleoresina Capsici. 1 drop. Powerful rubefacient.
" Cubebæ. 5–30♏.
" Filicis. ʒss.
" Lupulinæ. 2–5 grs.
" Piperis. 1–2♏.
" Zingiberis. 1♏, much diluted.

Resins.

1. *Resins proper.*— Resina, mastich, Copal, Elemi, Sandarac, Pix Canadensis, Pix Burgundica, Guaiaci resina, Succinum, Copaivic acid.
2. *Oleo resins.* — Terebinthina (white turpentine), Terebinthina Canadensis (balsam of fir), Terebinthina Veneta (Venice turpentine), Copaiba.
3. *Gum resins.*—Ammoniacum (stimulant, expectorant), Assafœtida (antispasmodic), Galbanum (stimulant, antispasmodic), Sagapenum (stimulant like assafœtida), Gambogia (acrid cathartic), Scammonium (cathartic), Olibanum (frankincense), Myrrha (emmenagogue and astringent).
4. *Balsams.* — Benzoinum, Balsamum Peruvianum (stimulant, expectorant), Balsamum Tolutanum (stimulant, expectorant), Styrax (stimulant and expectorant).
5. *Other articles containing resins or resinoid active prin-*

ciples. — Calamus, Cimicifuga, Colocynthis (colocynthin), Extract Cannabis (cannabin), Guaiaci lignin, Helleborus (Helleborin), Jalapa (jalapin), Mezereum, Podophyllum (podophyllin), Pyrethrum (pyrethrin), Zingiberis, and drugs containing essential oils. (See *Resinoid Extracts.*)

Neutral Organic Principles.

Names of alkaline principles terminate in *ia*, neutral or subacid principles in *in* or *ine*.

1. *Extractive matters, soluble in water.* — Aurantin (from cortex aurantii and limonis) — Bitter extractive of Anthemis, of Canella, of Chimaphila, of Coptis, of Cornus Florida, of Eupatorium, of Gentiana, of Marrubium, of Serpentaria, and acrid extractive of Scillæ — Cathartin (in senna, cassia, and rhamni bacca), Ergotin (extractive of ergot), Extractive of Juglans (cathartis), Ilicin (in ilex, used as substitute for quiniæ).

2. *Neutral crystalline principles.* — Absinthin (from absinthium), Aloin (from aloe), Amygdalin (from amygdala amara), Asparagin and Althein (from asparagus, althæa, glycyrrhizæ, and symphytum), Apocynin (from apocynum cannabinum, emetic and cathartic), Asclepione (from asclepias syriaca, narcotic), Caffein (from coffee, isomeric with thein), Cantharidin (from cantharis), Cascarillin (from cascarilla), Cetrarin (from cetraria), Columbin (from colomba), Cubebin (from cubeba), Cusparin (from angustura), Daphnin (from mezereum), Digitalin (from digitalis, a violent poison; dose, one-thirtieth of a grain), Elaterin (from elaterium, powerful cathartic; dose, one-tenth grain), Esculin (from æsculus), Helleborin (from helleboris), Hesperidin (from cortex limonis, etc.), Hydrastin (from hydrastis — see

page 90). Limonin (from seeds of lemon), Liriodendrin (from liriodendron), Maticin (from matico), Meconin, Narcein, and Narcotin (from opium), Phloridzin (from apple, cherry and plum trees), Picrotoxin (from cocculus indicus,) Piperin (from piper nigrum and longum), Quassin (from quassia and simaruba), Salicin (from salix, etc.), Santonin (from semen santonica — a popular anthelmintic — dose 1 to 4 grs.), Saponin (from saponaria), Sarsaparillin (from sarsaparilla), Scillitin (from scilla), Scoparin (from scoparius), Senegin (from senega), Taraxacin (from taraxacum), Thein (from tea — see Caffein), Theobromin (from theobroma cacao), Xanthoxylin (from xanthoxylum).

3. *Coloring principles.* — Indigotin (from indigo), Orcine (from litmus), Chlorophylle (from leaves), Carthamus (red and yellow), Carmine (from coccus), Polycroite (from crocus — yellow), Curcumin (from curcuma — yellow), Hæmatin (from hæmotoxylon), Quercitron (quercitron — yellow), Santalin (from santalum), Rubian, Alizarin, and Purpurin (from rubia), Anchusin (from anchusa — red, green), Rhabarbaric acid (from rheum — yellow, red, with alkali), Sanguinarina (from sanguinaria), Hydrastine (from hydrastis — yellow).

Vegetable Acids.

1. *Fruit acids.* — Citric (in lemon, etc.), Tartaric (in grapes, used in effervescing drinks), Malic (in apples, etc.), Oxalic (in rhubarb, sorrel, etc.), Pectic.
2. *Astringent and allied acids.* — Tannic (styptic, dose 2 to 10 grs.), Gallic (astringent, dose 5 to 20 grains), Ellagic, Catechu-tannic (in kino, catechu, etc.), Cincho-tannic (in cinchona bark), Cephælic (in ipecac.).

154 ORGANIC PRODUCTS.

3. *Balsamic, having relation to essential oils.* — Benzoic, Cinnamic, Valerianic and Hydrocyanic (sedative). Dose of officinal dilute acid ♏ii to v.
4. *Combined with alkaloids in plants.* — Meconic (with morphia, etc., in opium), Kinic (with quinia, etc., in cinchonas), Aconitic (with aconitia, etc., in aconite), Strychnic or Igasuric (with strychnia and brucia in nux vomica), Veratric (with veratria in cevadilla seed), Calumbic (with bebeerina, in columbo), Cevadic (with colchicia, in colchicum), Coccalinic (with menispermina, in cocculus indicus).

Vegetable Alkaloids.

Aconitia ($\frac{1}{2}$ to 2 grs. to ℨj of ointment in neuralgia), Delphina, Berberina, Cissampelina, Menispermina. The opium alkaloids are Morphia, Narcotin, Codeia, Paramorphia or Thebaina and Papaverina. Dose of morphia salts $\frac{1}{8}$ to $\frac{1}{4}$ gr. — Sanguinarina, Conia. The Cinchona alkaloids are Quinia, Quinidia, Cinchonia (Quinoidine or Chinoidine is precipitated extract), (Quinæ Sulphas is the well-known antiperiodic — the sulphates of the other alkaloids have been used as substitutes), Emetia or Emetina (active principle of ipecac), Arnicina, Lobelina, Strychnia (tonic excitomotor, dose $\frac{1}{12}$ gr.), Brucia (like strychnia, from nux vomica, etc., less powerful), Atropia (used in solution for dilating the pupil of the eye), Daturia, Hyoscyamia, Solania, Nicotia or Nicotina (active principle of tobacco), Bebeerina (the sulphate is tonic and antiperiodic, dose 3 to 10 grs.), Veratria (used in neuralgia and gout, dose $\frac{1}{12}$ gr., or externally in ointment, ℈j to ℨj), Colchicia.

III. Pharmaceutical Preparations.

Medicated Waters.

1st Class. — *By trituration with an insoluble substance, and subsequent filtration.*

Aqua Camphora. Camphor ʒj. Carb. Magnes. ʒij to Oj. Dose f℥ss.
" Amygdalæ Amaræ. Oil ♏xvj. Carb. Magnes. ʒj to Oij. Dose f℥j.
" Cinnamomi. Oil ♏xvj. Carb. Magnes. ʒj to Oj. Dose f℥ij.
" Fœniculi. Oil ♏xvj. Carb. Magnes. ʒj to Oj. Dose f℥ij.
" Menthæ Pip. Oil ♏xvj. Carb. Magnes. ʒj to Oj. Dose f℥ij.
" Menthæ Virid. Oil ♏xvj. Carb. Magnes. ʒj to Oj. Dose f℥ij.

2d Class. — *By distillation.*

Aq. Rosa. Rose petals ℔j to Oj.

3d Class. — *By charging water with gas.*

Aq. Acidi Carbonici, 5 parts CO_2 to 1 of water.

Infusions.

1st Class. — *With maceration, by boiling water.*

Group 1. — ℥j to Oj.

Infusum Cinchonæ Flavæ. Tonic.
" Cinchonæ Rubræ. Tonic.
" Cascarillæ. Stimulant, tonic.

156 PHARMACEUTICAL PREPARATIONS.

Infusum Eupatorii. Tonic — a diaphoretic, and emetic when hot.
" Krameriæ. Astringent.
" Sarsaparilla. Alterative, diaphoretic.
" Ulmi. Demulcent.
" Buchu. Demulcent, diuretic.
" Armoraciæ (with mustard-seed ʒj). Stimulant, diuretic.
" Senna (+ Coriander ʒj). Cathartic.

Group 2.— ℥ss. to Oj.

Infusum Angusturæ. Stimulant, tonic.
" Anthemidis. Tonic, emetic when hot.
" Calumbæ. Tonic.
" Serpentariæ. Tonic.
" Valerianæ. Stimulant, antispasmodic.
" Capsici. Arterial stimulant. Dose f℥ss.
" Zingiberis. Carminative.
" Humuli. Tonic, mild narcotic.
" Spigeliæ. Anthelmintic.
" Catechu Comp. (+ Cinnamon ʒj). Astringent.
" Lini Comp. (+ Liquorice Root ʒij). Demulcent.

Group 3.— Proportions varied.

Infusum Caryophylli ʒij to Oj. Stimulant.
" Rhei ʒij to Oj. Cathartic.
" Tabaci ʒj to Oj. Sedative injection in hernia.
" Digitalis ʒj to Oss. +Tinct. Cinnam. f℥j. Narcotic. Dose f℥j.
" Rosæ Comp. ℥ss. to Oijss. +Sugar, diluted Sulphuric Acid, Water. Adjuvant to astringent gargles.
" Taraxaci ℥ij to Oj. Diuretic.

2d *Class.* — *With cold water, by maceration or displacement.*

Infusum Cinchona Comp. ʒj to Oj. +Arom. Sulph. Acid fʒj. Tonic.
" Pruni Virginianæ ℥ss. to Oj. Sedative, tonic.
" Quassiæ ʒij to Oj. Tonic.
" Gentianæ Comp. ℥ss. to Oj. +Bitter Orange-peel, Coriander, Dil. Alcohol, Water. Tonic.
" Sassafras Medullæ ʒj to Oj. Demulcent.

Solutions or Liquors.

Liquor Ammoniæ Acetatis. Spiritus Mindereri. Spirit of Mindererus. Dilute Acetic Acid, saturated with Carbonate of Ammonia. A valuable diaphoretic. Dose ℥ss. to ℥jss.
" Ammoniæ Citratis. Dose fʒij to fʒvi.
" Arsenici et Hydrargyri Iodidi. Donovan's Solution. Alterative in skin disease. Dose 5 drops 3 times a day.
" Arsenici Hydrochloricus. Hydrochloric Solution of Arsenic. Similar to Fowler's Solution. Dose 2 to 8♏.
" Barii Chloridi. Solution of Chloride of Barium. Used in cancer and scrofula. Dose 5 drops.
" Bismuthi et Ammoniæ Citratis. Solution of Citrate of Bismuth and Ammonia. Dose fʒj.
" Calcii Chloridi. Solution of Chloride of Calcium. Dose 30♏ to ʒj.
" Calcis. Lime Water. Lime ℥iv. Water Oviii. Dose f℥ij to f℥iv during the day. The spray from the atomizer in membranous croup.
" Calcis Saccharatus. Saccharated Solution of Lime — ʒj equal to ℥j of Lime Water.
" Ferri Citratis. Solution of Citrate of Iron. Dose 10♏ equal to 5 grs. of Salt.
" Ferri Nitratis. Solution of Nitrate of Iron. Dose 10-15 drops.

Liquor Ferri Perchloridi Fortior. A powerful styptic.
" " Subsulphatis. Monsel's Solution. Styptic.
" Gutta-perchæ. Solution of Gutta-percha. (Sliced Gutta-percha ʒjss. Chloroform ʒxvii. Dissolve and add Carbonate of Lead ʒij. Settle and decant.) Artificial cuticle, etc.
" Hydrargyri Nitratis. Acid solution of Nitrate of Mercury. Caustic.
" Iodinii Compositus. Compound solution of Iodine. (Iodine 360 grs., Iod. Potass. ʒjss. Dist. Water Oj.) Dose 6 drops, diluted.
" Lithiæ Effervescens. Dose 5 to 10 f ʒ.
" Magnesiæ Carbonatis. Dose 1–2 f ʒ.
" " Citratis. Solution of Citrate of Magnesia. A pleasant cathartic. Dose ʒxii.
" Morphiæ Acetatis. Dose 15–30♏.
" " Hydrochloratis. Dose 15–30♏.
" " Sulphatis. (Sulph. Morph. 8 grs., Dist. Water Oss.) Dose fʒj equal to ⅛ gr. Magendie's Solution contains 16 grs. to f ʒj.
" Plumbi Subacetatis. Sedative and astringent lotion.
" Plumbi Subacetatis dilutus. fʒiij of Solution of Subacetate of Lead, to Oj Water.
" Potassæ. Antacid and Antilithic. Dose 10–30♏.
" " Arsenitis. Fowler's Solution. Average dose 10 drops, 2 or 3 times a day.
" " Citratis. Citric Acid ʒss. Bi-Carb. Potassa 330 grs. Water Oss.
" Sodæ Chlorinatæ. Labarraque's Disinfectant.

Tinctures.

Tinctures were formerly made by maceration for 2 weeks, but are now best prepared by percolation. The

Paris Society of Pharmacy direct percolation with Alcohol of 60° for Belladonna, Conium, Hyoscyamus, Stramonium, Digitalis, Quassia, Pale Cinchona, Rhetany, and Senna; with Alcohol of 80° for Valerian, Cinnamon, and Red and Yellow Cinchona. They advise maceration with Alcohol of 60° for Aloes, Arnica, Catechu, Colchicum bulbs, Gentian, Ipecacuanha, Musk, Rhubarb, and Squill. With Alcohol of 80° for Castor, Columbo, Colchicum seed, Cloves, Ginger, Jalap, Nux Vomica, Saffron, and Vanilla. With Alcohol of 90° for Assafœtida, Balsam of Peru, Balsam of Tolu, Benzoin, Ammoniæ, Myrrh, and Scammony.

Tinctura Aconiti Folii. Dose 20-30 drops.
" " Radicis. Dose 5-10 drops.
      Dr. Fleming's Tincture is stronger than this.
" Aloes. Purgative ℥ss to ℥j; Laxative ʒj-ʒij.
" " et Myrrhæ. Dose ʒj-ʒij.
" Arnica. A popular lotion.
" Assafœtidæ. Dose fʒss.-fʒj.
" Belladonnæ. 15-30 drops.
" Benzoini Compositæ. f℥ss.-f℥ij.
" Calumbæ. fʒj-f℥iv.
" Cannabis. 40 drops = 1 gr. of Extract.
" Cantharidis. 20 drops-fʒj.
" Capsici. fʒj-fʒij.
" Cardamomi. fʒj-fʒij.
" " Compositæ. fʒj-fʒij.
" Castorei. f℥ss.-f℥ij.
" Catechu. f℥ss.-f℥iij.
" Cinchonæ. fʒj-f℥iv.
" " Compositæ. fʒj-f℥iv.
" Cinnamomi. fʒj-f℥iv.
" Colchici. f℥ss.-f℥ij.
" Conii. f℥ss.-f℥j.
" Cubebæ. fʒj-fʒij.
" Digitalis. 10-20 drops.

160   PHARMACEUTICAL PREPARATIONS.

Tinctura Ferri Chloridi.   10 drops–f℥j.
" Gallæ.   f℥j–f℥iij.
" Gentianæ Compositæ.   f℥j–f℥ij.
" Guaiaci.   f℥j–f℥iij.
" " Ammoniata.   f℥j–f℥ij.
" Hellebori.   f℥ss.–f℥j.
" Humuli.   f℥j–f℥iij.
" Hyoscyami.   f℥ss–f℥i.
" Iodinii.   10–20 drops.
" " Composita.   15–30 drops.
" Jalapa.   f℥j–f℥ij.
" Kino.   f℥j–f℥ij.
" Krameriæ.   f℥j–f℥ij.
" Lobeliæ.   f℥j–f℥ij.
" Lupulinæ.   f℥j–f℥ij.
" Myrrhæ.   f℥ss.–f℥j.
" Nucis Vomicæ.   20♏.
" Opii.   25 drops = 1 gr.
" " Acetata.   20 drops = 1 gr.
" " Ammoniata.   30♏–f℥i.
" " Camphorata.   f℥j–f℥ij.   Infant 5–20 drops.
" " Deodorata.   25 drops.
" Quassiæ.   f℥j–f℥ij.
" Rhei.   f℥j–f℥ij.   As a purgative, f℥ss.–f℥j.
" " et Aloes.   Ditto.
" " et Gentianæ.   Ditto.
" " et Sennæ.   f℥ss.–f℥j
" Sanguinariæ.   30–60 drops.
" Scillæ.   20–40 drops.
" Serpentariæ.   f℥j–f℥ij.
" Stramonii.   20–40 drops.
" Tolutana.   f℥j–f℥ij.
" Valerianæ.   f℥j–f℥iv.
" " Ammoniata.   f℥ss.–f℥j.
" Veratri Viridis.   3–8 drops.

Tinctura Zingiber. 8–20♏.
    "   "   Fortior. 5–20♏.

Medicated Wines.

White or Sherry Wines used in making them.

Vinum Aloes ℨj+Cardamom. ⎰ to Oj. Dose fℨj to
            Ginger āā ℨj, ⎱ fℨij. Carminative, aperient.

" Rhei ℨiij+Canella ℨj, ⎰ to Oj. Dose fℨj to
   Diluted Alcohol fℨij, ⎱ fℨij. Carminative, aperient.

" Colchici Rad. ℨvj to Oj. Dose gtt. 10 to fℨj. Diuretic, nervous sedative.

" Colchici Sem. ℨij to Oj. Dose fℨss. to fℨij. Diuretic and nervous sedative.

" Ergotæ ℨij to Oj. Dose fℨj. Excito-motor stimulant. Dose ℨj to ℨiij.

" Ipecacuanhæ ℨj to Oj. Dose fℨj to fℨss. Expectorant, 10–30♏.

" Tabaci ℨj to Oj. Dose gtt. 20. Diuretic.

" Antimonii, 2 grains Tartar Emetic to fℨj. Dose fℨj to fℨss. Expectorant, emetic.

" Opii ℨij+Cinnamon. Cloves āā 60 grs. to Oj. Dose same as Tincture. 2 or 3 drops as Collyrium.

Vinegars.

With 1 part Acetic Acid to 7 of water.

Acetum Colchici ℨj to Oj, Alcohol fℨss. Dose gtt. ♏30 to fℨij. Diuretic, sedative.

" Scillæ ℨij to Oj. Dose gtt. 30 to fℨj. Diuretic, sedative.

" Opii ℨijss. to Oj. Dose gtt. 5 to 10. See p. 162.

Preparations of Opium.

Tinct. Opii Camphorata (Paregoric) { Opium ʒss., Camphor Ðj, Benzoic Acid ʒss., Oil Anise fʒss., Honey ʒj, } to Oj Diluted Alcohol. Dose fʒj to fʒss.

" Opii (Laudanum) Opium ʒx. to Oj. Dose gtt. 25.

" Opii Acetata { Opium ʒj, Alcohol fʒiv, Vinegar fʒvj. } Dose gtt. 20.

Vinum Opii (Sydenham's Laud.) { Opium ʒij, Cinnamon, Cloves āā ʒj, } to Sherry Oj. Dose gtt. 20.

Acetum Opii (Black drop) { Opium ʒv, Nutmeg ʒj, Sugar ʒviii. } to Oii. Dose gtt. 5 to 10.

Liquor Morphia Sulphatis, ⅛ grain Morphia to fʒj. Dose fʒj. (Magendie's Solution, used in New York and Boston, has 16 grains to fʒj.)

Decoctions.

Decoctum Chimaphilæ ʒj to Ojss.; boil to Oj. Dose Oj per diem. Alterative.
" Uvæ Ursi ʒj to fʒxx; boil to Oj. Dose fʒij. Astringent, diuretic.
" Dulcamaræ ʒj to Ojss.; boil to Oj. Dose fʒij. Alterative, narcotic.
" Hæmatoxyli ʒj to Oij; boil to Oj. Dose fʒj. Astringent.
" Quercus Alb. ʒj to Ojss.; boil to Oj. Dose fʒij. Astringent.
" Cinch. flav. ʒj to Oj; boil ten minutes. Dose fʒij. Tonic.

Decoctum Cinch. rub. ʒj to Oj; boil ten minutes. Dose f ʒij. Tonic.
" Cornus Floridæ ʒj to Oj; boil ten minutes. Dose f ʒij. Tonic.
" Senega ʒj to Ojss.; boil to Oj. Dose f ʒij. Stimulant, expectorant.
" Hordei ʒij to Oivss.; boil to Oij. Ad libitum. Demulcent.
" Cetrariæ ʒss. to Ojss.; boil to Oj. Dose Oj per diem. Tonic, demulcent.
" Taraxaci ʒij to Oij; boil ten minutes. Dose f ʒij. Diuretic.
" Sarsap. Comp.:
  Sarsap. ʒvj,
  Sassafras,
  Guaiac,
  Liquorice āā ʒj,
  Mezereon ʒiij,
  } to Oiv; boil fifteen minutes. Dose f ʒiv. Alterative, diaphoretic.

Mixtures.

Mistura Ammoniaci. 240 grs. Ammoniac, Oj Water. Dose 1–2 teaspoonfuls.
" Amygdalæ. Almond mixture. Sweet Almond ʒss., Gum Arabic 30 grs., Sugar 120 grs., Water ʒviii. Demulcent.
" Assafœtidæ. Assafœtida 120 grs., Water Oss. f ʒij–f ʒiv as enema.
" Chloroformi. Chloroform ʒss. (by weight), Camphor 60 grs., Water f ʒvi. Suspend by yolk of egg. Dose 1–2 tablespoonfuls.
" Creasoti. Creasote. Glacial Acetic Acid āā 16♏, Spts. Juniper f ʒss., Syrup f ʒj, Water f ʒxv. Dose f ʒj.
" Cretæ. Chalk mixture. Prepared chalk ʒss., Sugar, Gum Arabic āā 120 grs., Cinnamon water, Water āā f ʒiv. Dose 1 tablespoonful.

Mistura Ferri Composita. Myrrh, Sugar āā 60 grains, Carb. Pot. 25 grs., Sulphate of Iron 20 grs., Spts. Lavender f℥ss., Rose Water f℥viiss. Dose f℥j–f℥ij.

" Glycyrrhizæ Composita. Brown Mixture. Liquorice extract, Sugar, Gum Arabic āā ℥ss., Tinct. Opii Camph., f℥ij, Vinum Antimonii f℥j, Spts. Eth. Nit. f℥ss., Water f℥xii. Dose for adult 1 tablespoonful.

" Potassæ Citratis. Neutral Mixture. Fresh lemon-juice Oss., Bicarb. Pot. q. s. Dose 1 tablespoonful.

Extracts.

1st Class. — *Narcotic inspissated juices.*

Extractum Aconiti. Dose ½ to 1 gr.
" Belladonnæ. Dose ¼ to 1 gr.
" Stramonii fol. Dose 1 to 2 grs.
" Conii. Dose 2 to 3 grs.
" Hyoscyami. Dose 2 to 3 grs.

2d Class. — *Hydro-alcoholic and alcoholic extracts.*

Extract. Aconiti Alcoholicum. ½ gr. to 1 gr. Narcotic.
" Arnicæ Alc. 5–10 grs.
" Belladonnæ Alcoholicum. ¼ gr. to ½ gr. Narcotic.
" Cannabis Purificatum. ¼ to 1 gr. Powerful narcotic.
" Stramonii Alcoholicum. ½ to 1 gr. Narcotic.
" Digitalis Alc. ¼ gr. Sedative, diuretic.
" Conii Alc. 1 to 2 grs. Narcotic.
" Hellebori Alc. 5–10 grs. Drastic, cathartic.
" Hyoscyami Alc. 1 to 2 grs. Narcotic.

PHARMACEUTICAL PREPARATIONS. 165

Extract. Nucis Vomicæ Alc. ½ gr. to 1 gr. Nerve stimulant.
" Ignatiæ Alc. ½-1 gr. Excito-motor tonic.
" Jalapæ. 10-15 grs. Cathartic.
" Rhei. 10-15 grs. Cathartic.
" Podophylli. 5-10 grs. Cathartic.
" Dulcamara Alc. 3-6 grs. Alterative, narcotic.
" Cinchonæ. 10-15 grs. Tonic, alterative.
" Sarsaparilla. 10-15 grs. Tonic, alterative.
" Colocynthidis Comp. 5-30 grs. Cathartic.

3d Class. — By displacement with cold water and evaporation.

Extract. Gentianæ. 10-20 grs. Tonic.
" Quassiæ. 3-6 grs. Tonic.
" Krameriæ. 10-20 grs. Astringent.
" Juglandis. 10-20 grs. Cathartic.
" Opii. Dose 1 gr. Narcotic.

4th Class.

Extract. Hæmatoxyli. 10-20 grs. Astringent. (Decoction evaporated.)
" Taraxaci. ℈j to ʒj. Diuretic. (Evaporated juice.)
" Colchici Aceticum. 1-3 grs. Diuretic. (Medicated vinegar evaporated.)
" Glycyrrhizæ. Liquorice. Demulcent.

Concentrated or Resinoid Extracts.

Resinæ Jalapæ. 2-5 grs. Cathartic.
" Podophylli. ¼-1 gr. Cathartic.
" Scammonii. 4-8 grs. Cathartic.
Cimicifugin. 1-6 grs.
Stillingin. 2-4 grs.

166 PHARMACEUTICAL PREPARATIONS.

Leptandrin. 2–4 grs.
Hydrastin. 1–2 grs.

Fluid Extracts.

1. *Syrups.*

Extract. Spigeliæ Fluidum. f ℨj. Cathartic.
" Rhei Fluidum. 20♏. Cathartic.
" Sennæ Fl. f ℨj.
" Sarsaparillæ Fl. 30–60♏. Diaphoretic, etc.
" Sarsaparillæ Comp. Fl. 30♏–f ℨj.
" Cinchonæ Fl. f ℨj.
" Uvæ Ursi Fl. 30♏–f ℨj.
" Dulcamaræ Fl. f ℨss.
" Pruni Virginianæ Fl. f ℨj–f ℨij.

2. *Alcoholic.*

Extract. Valerianæ Fluidum. f ℨj. Antispasmodic.
" Serpentariæ Fl. 20–30♏.
" Buchu Fl. 20–30♏.
" Conii Fl. 4–5♏.
" Ergotæ Fl. 10–20♏.
" Ipecacuanhæ Fl. 15–30♏.
" Hyoscyami Fl. 5–10♏.
" Taraxaci Fl. f ℨj.
" Gentianæ Fl. 10–40♏.
" Zingiberis Fl. 10–20♏.
" Lupulinæ Fl. 10–15♏.
" Opii liquidum. 10♏.
" Veratri Viridis Fl. 2–4♏.

Syrups.

Syrups are: 1. Infusions or decoctions made permanent by sugar.

PHARMACEUTICAL PREPARATIONS. 167

2. Sugar added to alcoholic extract.
3. Sugar added to acetic extract.
4. Additions to simple syrup.

Syrupus. Simple Syrup. 18 oz. Sugar to Oj. ℥vi.
" Acaciæ. Syrup of Gum Arabic. Demulcent.
" Acidi Citrici. Refrigerant.
" Allii. Nerve stimulant. Dose ʒj.
" Amygdalæ. Sedative, demulcent.
" Aurantii Corticis. Flavoring.
" " Florum. Flavoring. Dose ʒj.
" Ferri Iodidi. Dose 20–40♏ diluted.
" Ipecacuanhæ. Emetic. 30♏ to ʒj for a child, ʒiv to ℥j to adult. Expectorant 2♏–10♏ for a child, ʒss–ʒj for adult.
" Krameriæ. ℥ss. for adult. 20–30♏ for a child.
" Lactucarii. Dose ʒij–ʒiij.
" Limonis. Flavoring.
" Papaveris. Syrup of Poppies. ʒss. to ʒj for an infant.
" Pruni Virginianæ. Dose f℥ss.
" Rhei. Mild cathartic for children. Dose ʒj.
" " Aromaticus. Same as last. Dose ʒj.
" Rosæ Gallicæ. Dose ʒj.
" Rubi. Syrup of Blackberry Root. Dose ʒj–ʒij.
" Sarsaparillæ Compositus. Dose ℥ss.
" Scilla. Expectorant. ʒj.
" " Compositus. Hive Syrup. Dose for children 10 drops–ʒj.
" Senegæ. Dose ʒj to ʒij.
" Tolutanus. Flavoring.
" Zingiberis. Dose ʒj or more.

Troches.

Trochisci Acidi Tannici. ½ gr. each.
" Bismuthi. 2 grs. each.

Trochisci Catechu. 1 gr. each.
" Cretæ. Antacid, astringent.
" Cubebæ. 1 drop oleoresin of cubeb each.
" Ferri Subcarbonatis. Each about 5 grs.
" Glycyrrhizæ et Opii. $\frac{1}{10}$ gr. opium each.
" Ipecacuanhæ. $\frac{1}{4}$ gr. each.
" Magnesia. Antacid.
" Menthæ Piperitæ. Carminative.
" Potassæ Chloratis. 5 grs. each.
" Sodæ Bicarbonatis. 5 grs. each.
" Zingiberis. Each 1–2♏ of tincture.

Confections.

Confectio Rosæ. Powd. Rose 2 pts., Sugar 15 pts., Honey 3 pts., Rose-water 4 pts.
" Aromaticæ. Arom. Powd., Honey equal parts.
" Opii (1 gr. in 36). Opium Powd. 270 grs., Arom. Powd. ℨvi, Honey ℨxiv.
" Sennæ. P. Senna and Coriander, with Figs, Prunes, Cassia, etc.

Jellies.

Convenient method of giving Cod-liver oil, Castor oil, Balsams, etc. Made of honey, syrup, gum Arabic, isinglass. The latter in solution added to an emulsion of the others.

Spirits, or Essences.

Spiritus Chloroformi. One Troyounce Chloroform, Diluted Alcohol, f℥xii. Dose ʒss. to ʒj.
" Cinnamomi. Dose 10–20 drops.
" Juniperi Compositus. Dose ʒij–ʒiv.
" Lavandulæ. Dose 30♏ to fʒj.

Spiritus Lavandulæ Compositus. Dose 30 drops to ʒj.
" Limonis. Flavoring.
" Menthæ Piperitæ. Dose 20-30 drops.
" " Viridis. Dose 30-40 drops.
" Myristicæ. Dose ʒj-ʒij.
" Myrciæ. Bay-rum. Perfume.
" Ætheris Compositus. Hoffman's Anodyne. Stimulating antispasmodic. Dose ʒss-ʒj.
" Ætheris Nitrosi. Sweet Spirit of Nitre. Diaphoretic, diuretic, antispasmodic. Dose ʒss.-ʒj.
" Ammoniæ. Stimulant, antispasmodic. Dose 10-30 drops.
" Ammoniæ Aromaticus. Same as last. Dose ʒss.-ʒj.
" Anisi. Carminative. Dose ʒj-ʒij.
" Camphoræ. Dose 5 drops to ʒj. Anodyne lotion.

Powders.

Medicines adapted to the form of powders are insoluble mineral substances, vegetable products, and some soluble substances.

Insoluble—too large doses for pills.—Carbo-ligni, Magnesia, Calcis Phosph., Pot. Bitart., Sulphur Sub., Creta Ppt., Ferri Subcarb., Calomel, etc. Vegetable powders, as Powd. Cinchona, Colomba, Gentian, Rhubarb, Jalap, Cubebs, etc.

In certain combinations, and when pills are objected to.— Powd. pil. Hydrarg., Ext. Colocy., Opium, Digitalis, Nux Vom., Kino, Tannic Acid, Gallic Acid, Potas. Nit., Opium Alkaloids, Cinchona Alkaloids, Subnit. Bismuth, etc.

Diluents for Powders.—Sugar, Lactin, Powd. Acacia, Cinnamon, Arom. Powd., Ext. Liquorice, Gum Tragacanth, Elm Bark, etc.

170   PHARMACEUTICAL PREPARATIONS.

Pulveres Effervescentes. Effervescing powders. 25 grs. Tartaric Acid in one paper, and 30 Bicarb. Soda in another. Mix in Oss. of water, with ginger, syrup, etc.

" Effervescentes Aperientes. Seidlitz powders. 35 grs. Tartaric Acid in one paper and 40 grs. Bicarb. Soda with ℨij Rochelle salt in the other. Used as the above, in Oj water.

Pulvis Aloes cum Canella. Hiera Picra. Aloes ℥xii, Canella ℨiij. Dose 10–20 grs.

" Aromaticus (Comp. Powder of Cinnamon). Cinnamon. Ginger, āā ℨij., Cardamom, Nutmeg, āā ℨj. Dose 10–30 grs.

" Ipecacuanhæ Compositæ. Dover's Powder. Ipecac., Opium, āā 60 grs., Sulph. Potas., ℨj. Dose 5–15 grs.

" Jalapæ Compositus. Jalap ℨj, Bitart. Potass. ℥ii. Dose 30 grs. to ℨj.

" Rhei Compositus. Rhubarb ℥iv, Magnesia ℥xii, Ginger ℥ij. Dose ℨss.–ℨj, a child 2 years old 5–10 grs.

Pills.

Medicines adapted to the pilular form are, powders in less than 15 grain doses, gum-resins, extracts, and oleoresins and oils in small proportions.

Unadhesive materials. — Calomel, Dover's Powder, Subnit. Bismuth, Morphia Acetas, etc., Strychnia, Pulv. Digitalis, Pulv. Ipecac., Plumbi Acetas, Ant. et Pot. Tartrate, Ant. Sulph., Argenti Nitras, Argenti Oxidum, Ferri Pulvis., Ferri Subcarbonas, etc., Potas. Iodid., Camphor, etc.

Medicinal excipients. — Extracts, Pil. Hydrarg., Pil. Copaibæ, Pil. Ferri Carb., Terebinthina.

With moisture. — Powd. Aloes, Rheum, Kino, Tannin, Opium, Scilla, Ferri Citras, Assafœtida, etc.

PHARMACEUTICAL PREPARATIONS. 171

With alcohol. — Guaiacum, etc.
With dilute SO_3. — Quiniæ Sulph., Cinchoniæ Sulph., Quinidiæ Sulph., Quinoidine.
Inert excipients. — Powd. Acacia, Tragacanth, Soap, Bread Crumbs, Confections, Syrup of Gum, Honey, Molasses, Syrups.

Pilulæ Aloes.   Socotrine Aloes, Soap, equal parts. Dose 5 pills as a purge.
" Aloes et Assafœtidæ. Aloes, Assafœtida, Soap, eq. pts. Dose 2–5 pills.
" Aloes et Ferri. Sulphate of Iron ʒjss., Aloes ʒij, Comp. Cinnamon powder ʒiij, Confection of Roses ʒiv. Dose 5–10 grs.
" Aloes et Mastiches. Aloes ʒjss., Mastic, Red Rose āā ʒss. Make 400 pills. Dose one after each meal as laxative.
" Aloes et Myrrhæ. Aloes ʒij, Myrrh ʒj, Aromatic powder ʒss., Syrup q. s. = 480 pills. Dose 10–20 grains.
" Antimonii Compositæ. Compound Calomel pill. Plummer's pill. Sulphuretted Antimony, Mild Chloride of Mercury, āā 120 grs. Guaiac, Molasses, āā ʒss. = 240 pills. Dose 1–2 pills.
" Assafœtidæ. Assafœtida ʒjss. Soap ʒss. = 240 pills. Dose 1–3.
" Cambogiæ Compositæ. Gamboge, Aloes, Compound powder of Cinnamon āā ʒj, Soap ʒij, Syrup q. s. Dose 5–15 grs.
" Catharticæ Compositæ. Compound Extract of Colocynth ʒss., Ext. Jalap, Calomel, āā 180 grs., Gamboge 40 grs. = 180 pills. Dose 3.
" Copaibæ. Copaiba ʒij, Magnesia 60 grs. = 200 pills. Dose 2–6.
" Ferri Carbonatis — Vallet's pills. — Sulphate of Iron ʒviii, Carb. Soda ʒix, Honey ʒiij, Sugar

℥ij, Boiling Water Oij, Syrup q. s. Reduced to ℥viii. Dose 10–30 grs. a day.

Pilulæ Ferri Compositæ. Myrrh 120 grs., Carb. Soda, Sulph. Iron, āā 60 grs, Syrup q. s. = 80 pills. Dose 2–6, 3 times a day.

" Ferri Iodidi. Iodine ℥ss., Iron 120 grs., Sugar ℥j, Marshmallow ℥ss., Gum Arabic, reduced Iron, āā 60 grs., Water f℥x. = 300 pills — 1 gr. Iodide Iron and ⅕ gr. of reduced Iron in each pill.

" Galbani Compositæ. Galbanum, Myrrh āā 360 grs., Assafœtida 120 grs., Syrup q. s. = 240 pills. Dose 5–20 grs.

" Hydrargyri — Blue pill. Mercury ℥j, Rose Confection ℥jss., Liquorice powder ℥ss. = 480 pills. Dose 5–15 grs. as cathartic, 1 two or three times a day as alterative.

" Opii. Opium 60 grs., Soap 12 grs. = 60 pills. Dose 1 pill.

" Quiniæ Sulphatis. Sulph. Quinia ℥j, Gum Arabic 120 grs., Honey q. s. = 480 pills. Dose 1 gr. of Sulph Quin. to each pill.

" Rhei. Rhubarb 360 grs., Soap 120 grs. = 120 pills. Dose 3 grs. to a pill.

" " Compositæ. Rhubarb ℥j, Aloes 360 grs., Myrrh ℥ss., Oil Peppermint f℥ss. = 240 pills. Dose 2–4.

" Saponis Compositæ. Opium 60 grs., Soap ℥ss. Dose 1 gr. Opium in 5 grs.

" Scillæ Compositæ. Squill 60 grs., Ginger, Ammoniac, āā 120 grs., Soap 180 grs., Syrup q. s. = 120 pills. Dose 5–10 grs.

Liquids.

Suitable medicines are most soluble salts, light insoluble powders, extracts, gum-resins, oils, etc.

PHARMACEUTICAL PREPARATIONS. 173

Forming eligible solutions with water. — Alumen, Ammon. Murias., Ant. et Pot. Tart., Barrii Chloridum, Calcii Chloridum, Ferri Sulph., Ferri et Pot. Tart., Manganesii Sulph., Magnesiæ Sulphas, Potassæ Acetas, Pot. Bicarb., Pot. Carb., Pot. Citras, Pot. Chloras, Pot. Tartras., Potassii Bromid., Pot. Iodid., Morphia Acetas, Morphia Sulphas, Morphia Murias, Sodæ Bicarb., Sod. Boras, Sod. Carbonas, Sod. Sulph., Sod et Pot. Tart., Sod. Chlorid., Sod. Phosph., Acid Citric, Acid Tartaric, Acid Tannic.

Mixing, but not forming clear solutions in water diffused by agitation. — Magnesia, Potas. Bitart., Sulph. Præcip., Pulv. Cinchonæ, Pulv. Ipecac., Quiniæ Sulph.

Miscible by trituration alone. — Ext. Aconiti, Ext. Bellad., Ext. Conii, Ext. Hyoscyami, Ext. Stramon., Ext. Taraxaci, Ext. Kramer., Ext. Glycyrrh., Confections, Assafœtida, Ammoniac, Guaiacum, Myrrha, Scammony.

Suspended by viscid excipients. — Copaiba, Ol. Amygdalæ, Ol. Ricini, Ol. Terebinth, Olea Essentia, Ferri Protocarb.

Requiring additions to form solutions. — Quin. Sulph., Cinchon. Sulph., Quinid. Sulph., Chinoidine, Iodine, Hydrarg. Iodid. Rub.

Requiring viscid substances. — Ammon. Carb., Hyd. Chlorid. Corros., Pot. Cyanuret, Potassa.

Vehicles or correctives (especially of salines). — Aq. Medicatæ, Syrups, Tinct. Cinnamomi, Tinct. Cinnam. Comp., Tinct. Cardam.

Tinct. Card. Com., Infus. Rosa Comp., Saccharum, Olea destillata, Tinct. Tolutana, Tinct. Zingiberis.

Glycerates.

Glyceritum Acid Carbolici. Glycerate of Carbolic Acid. Glycerine 4 parts, Carbolic Acid 1 part — $4\frac{1}{2}$ m, represent 1 gr. of Acid. Dose 5–10 m.

174 PHARMACEUTICAL PREPARATIONS.

Glyceritum Acidi Gallici. Glycerate of Gallic Acid. Glycerine 4 pts., Gallic Acid 1 pt. Dose 20–60♏.
" Acidi Tannici. Glycerate of Tannic Acid. Glycerine 4 pts., Tannin 1 pt. Dose 10–40♏.
" Boracis. Glycerate of Borax. Glycerine 4 pts., Borax 1 part. Convenient application to aphthæ.

Liniments.

Linimentum Aconiti. ℥xvi to Oj Rect. Sp.+℥j Camphor.
" Ammoniæ. Aq. Amm. ℥j, Olive Oil ℥ij.
" Belladonnæ. See Lin. Acon.
" Camphoræ. Camph. ℥ijj, Olive Oil ℥iv.
" " Comp. Camph. ℥ijss., Ol. Lavend. ʒj, Aq. Ammon. ℥v, Rect. Spirit ℥xv.
" Cantharidis. Canth. ℥j, Ol. Terebinth. Oss.
" Chloroformi. Chloroform ℥iij, Olive Oil ℥iv.
" Crotonis. Croton Oil ℥j, Ol. Cajeput. Rect. Spt. āā ℥iijss.
" Hydrargyri. Ung. Hyd. ℥j, Aq. Am., Lin. Camph. āā ℥j.
" Iodi. Iodine ʒx, Iod. Pot. ℥ss., Camph. ʒij, Rect. Sp. ℥x.
" Opii. Tr. Opii, Lin. Sap. āā ℥ij.
" Potassii Iodidi cum Sapone. Soap, Iod. Pot. āā ℥jss., Glycerine ℥j, Ol. Limonis ℥j, Water ℥x.
" Saponis. Soap ℥iv, Camph. ℥ij, Ol. Rosmar. ℥ss., Water ℥iv, Alcohol Oij.
" Sinapis Comp. Ol. Mustard ʒj, Ext. Mezereon 40 grs., Camphor 120 grs., Castor Oil ʒv, Rect. Sp. ℥iv.

Linimentum Terebinthinæ. Resin cerate ℥xii, Ol. Tereb. Oss.
"   "   Aceticum. Ol. Tereb., Acetic Acid, Lin. Camp. āā ℥j.

Cerates and Ointments.

1st Class. — *Much used as Vehicles.*

Ceratum Adipis or Simplex. 1 pt. White Wax, 2 Lard. Firmest healing dressing.
"   Cetacei. 1 Cet., 3 White Wax, 6 Olive Oil. Firm healing dressing.
Unguentum Simplex. 1 White Wax, 4 Lard. Softer healing dressing.
Ung. Aquæ Rosæ. { Almond Oil, Spermaceti, White Wax, Rose Water. } Softest healing dressing.
Ceratum Resinæ. { 5 Resin, 8 Lard, 2 Yellow Wax. } Stimulant healing dressing. (Basilicon.)

2d Class. — *Mechanical mixture of medicinal substance with unctuous ingredient.*

Group 1. — By fusion.

Cerat. Resinæ Comp. { Resin, Suet, Yellow Wax, Turpentine, Flaxseed Oil. } Stimulating.
Ung. Picis Liq. Tar and Suet, equal parts. Stimulant, antiseptic.
Cerat. Cantharidis. { Canth. 12 pts., Lard 10. Yel. Wax, Resin āā 7 pts. } Epispastic, blist. cerat.

Group 2. — By trituration.

Cerat. Sabinæ. 1 pt. Savin, 6 Resin Cerate. Stimulant dressing.
Ung. Gallæ. 1 pt. Galls, 7 Lard. Astringent.
Ung. Veratri alb. 1 pt. Root, 4 Lard and Oil Lemon. In itch.

PHARMACEUTICAL PREPARATIONS.

Cerat. Calaminæ. $\{$ ℨiij Calamine, ℥xij Lard, ℨiij Wax. $\}$ Mild ast. and desiccant.

Cerat. Zinci Carb. $\{$ 1 pt. ZnO,CO_2, 5 Simple Ointment. $\}$ Mild astringent.

Ung. Zinci Oxidi. $\{$ 1 pt. ZnO, 6 Lard. $\}$ Mild astringent.

Ung. Cupri Subacet. $\{$ 1 pt. $2CuO,\overline{A}c,6HO$, 15 pts. Simple Ointment. $\}$ Mild escharotic.

Ung. Antimonii. $\{$ 1 pt. KO,SbO_3,T, 4 pts. Lard. $\}$ Vesicant.

Ung. Hydrargyri. Equal parts Hg and Lard. Alterative.

Ung. Hydrargyri Ammon. $\{$ 1 pt. Hg_2Cl,NH_2, 12 Simple Ointment, $\}$ Desiccant, alterative.

Ung. Hydrargyri Iodidi Rub. $\{$ HgI, 16 grs. to ℥j, Simple Ointment, $\}$ Stimulant, alterative.

Ung. Iodinii. $\{$ ƌj I, 4 grs. KI, Lard ℥j. $\}$ Discutient, alterative.

Ung. Iodini Comp. $\{$ 1 pt. I, 2 pts. KI, 32 Lard. $\}$ Discutient, alterative.

Ung. Potassii Iodid. $\{$ 1 pt. KI+1 pt. Aq., 8 Lard. $\}$ Discutient, alterative.

Ung. Plumbi Carb. $\{$ 1 pt. PbO,CO_2, 6 Ung. Simp. $\}$ Astringent, desiccant.

Ung. Sulphuris. 1 pt. S, 2 pts. Lard. In Itch.

Ung. Sulphuris Comp. $\{$ Sulph. ℥j, Ammon. Merc. ʒj, Benz. Acid ʒj, Ol. Berg, fʒj, Sulph. Acid fʒj, Nit. Potas. ʒij, Lard ℥vj. $\}$ In Itch.

Ung. Belladonnæ. 1 pt. Ext., 8 Lard. Anodyne.
Ung. Stramonii. 1 pt. Ext., 8 Lard. Anodyne.
Ung. Creasoti. Creasote fʒss., Lard ℥j. Antiseptic, stimulant.
Ung. Benzoini. Tr. Benz. ʒj to ℥iv Lard. Soothing.
Ung. Cadmii Iodidi. 62 grs. to ℥j. Alterative.
Ung. Plumbi Iod. 62 grs. to ℥j. Alterative.

PHARMACEUTICAL PREPARATIONS. 177

3d Class. — *By digesting the ingredient in Lard.*

Ung. Tabaci. ℥j Leaves to lb.j Lard. Narcotic.
Ung. Mezerei. ℥iv Bark to ℥xiv Lard, ℥ij Wax. Stimulant.
Ung. Cantharidis (with boiling water), ℥ij to ℥viij. Resin Cerate. Stimulant.
Ung. Picis Liquidæ. Equal parts Tar and Suet. Stimulant.
Ung. Zinci Oxidi. 80 grs. to ℥j Ung. Benz. Mild Astringent.

4th Class. — *The unctuous ingredient is decomposed.*

Ung. Hydrarg. Nit. (Citrine Ointment.) Powerful stimulant and alterative.
Cerat. Saponis. Soothing dressing.
Cerate Plumbi S. Acet. (Goulard's Cerate.) Cooling.

Plasters.

Emplastrum Ammoniaci. { Ammoniac ℥v, Dil. Acetic Acid Oss. Dissolve, strain, and evaporate. } Stimulant.

Emp. Ammoniaci cum Hydrargyro. { Ammoniac ℥xii, Mercury ℥iii. Olive Oil 60 grs., Sublimed Sulphur 8 grs. } Stimulant, alterative.

Emp. Antimonii. Tartar Emetic ℥j, Burgundy Pitch ℥iv. Pustulating.

Emp. Arnicæ. Alcoholic Ext. Arnica ℥jss. Resin Plaster ℥iij. Stimulant.

Emp. Assafœtidæ. { Assafœt., Lead Plaster, āā ℥xii, Galbanum, Yellow Wax, āā ℥vi, Alcohol Oiij. Dissolve the gums in Alcohol, evaporate till like honey, melt all together, stir, evaporate. } Stimulant. Antispasmodic.

12

178 PHARMACEUTICAL PREPARATIONS.

Emp. Belladonnæ. Alcoholic Ext. Bellad. ℨj, Resin Plaster ℨij. Anodyne.

Emp. Ferri. { Sub. Carb. Iron ℨiij, Lead plaster ℨxxiv, Burgundy Pitch ℨvi. } "Strengthening plaster."

Emp. Galbani Comp. { Galbanum ℨviii, Turpentine ℨj, Burg. Pitch ℨiij, Lead Plaster ℨxxxvi. } Stimulant.

Emp. Hydrargyri. { Mercury ℨvi, Olive Oil, Resin, āā ℨij, Lead Plaster ℨxii. } Alterative.

Emp. Opii. Ext. Opium ℨj, Burg. Pitch ℨijj, Lead Plaster ℨxii. Anodyne.

Emp. Picis Burgundicæ. Burg. Pitch ℨ72, Yellow Wax ℨ6. Rubefacient.

Emp. Picis cum Cantharidæ. Burg. Pitch ℨ48, Canth. cerate ℨ4. Rubefacient.

Emp. Plumbi. Litharge ℨ30, Olive Oil ℨ56. "Lead Plaster," "Diachylon."

Emp. Plumbi Iodidi. Iod. Lead ℨj, Soap Plaster, Resin Plaster, āā ℨiv. Alterative.

Emp. Resinæ. Resin ℨvi, Lead Plaster ℨ36. Stimulant.

Emp. Saponis. { Hard Soap ℨ10, Yellow Wax ℨ$12\frac{1}{2}$, Olive Oil Oj, Oxide of Lead ℨ15, Vinegar 1 gal. Melt, evaporate. } "Soap Plaster."

Suppositories.

Suppositoria Acidi Tannici. 3 grs. to each.
" Hydrargyri. 5 grs. to each.
" Morphiæ. $\frac{1}{2}$ gr. to each.
" Plumbi Compositæ. 3 grs. Acetate Lead and 1 gr. Opium to each.

VIII.

TABLE OF SYMPTOMATOLOGY.

A. TOPOGRAPHICAL.

1. — General Aspect of Patient.

I. Attitude.

Unusual languor — invasion of acute disease — course of chronic ones.
Unnatural boldness — insanity — delirium.
General immobility — catalepsy.
Irregular and perpetual movement — chorea.
Distorted features, altered position, and impaired motion of limbs — hemiplegia.
Tonic spasm of trunk — tetanus.

II. Decubitus.

Constantly dorsal — cerebral apoplexy, organic disease of brain and spinal marrow, acute peritonitis, general articular rheumatism.
Prone — generally in gastric, intestinal, hepatic, and renal colic.
Lateral — some stages of pleurisy or pneumonia (not

general); in consumption, when one lung is affected, the diseased side is usually lain on.

Sitting — diseases of heart and lungs, which interfere with respiration.

Head thrown back — laryngeal and tracheal disease.

Restlessness, jactitation, etc. — the invasion of acute inflammation, idiopathic fevers, many affections of children, delirium and acute mania.

III. Volume of Body.

General enlargement — anasarca, or emphysema from a wound of the chest.

2. Signs furnished by Head, Face, and Neck.

Head bent to one side — convulsions, hemiplegia, torticollis, dislocation of cervical vertebra, cervical glandular swellings, cicatrices of neck, after burns.

Head bent forward — vertebral malformation.

Head bent back — diseases with dyspnœa, as croup, laryngismus stridulus, suffocative catarrh, etc. Tetanus, spinal meningitis of neck.

Cranium increased in size — chronic hydrocephalus — hypertrophied brain.

Œdematous scalp — erysipelas, small-pox.

Facies stupida (dull expression) — typhoid fever.

Facies vultuosa (full, red face, injected eyes) — cardiac hypertrophy, cerebral congestion.

Pinched countenance (opposite of last) — acute peritonitis, in health, from exposure to severe cold.

Facies hippocratica — in chronic diseases just before death, in unusually prolonged acute disease.

3. Physiognomical Rugæ.

Rugæ transversæ (in forehead) — excessive pain rising externally.

R. oculo-frontales (from forehead vertically to root of nose) — distress, anxiety, anguish, and severe internal pain. In acute diseases, an imperfect or false crisis, impending efflorescence, and often fatal termination. Linea oculo-zygomatica (from inner angle of the eye below the cheek-bone) — in children a cerebral or nervous affection; in adults, disorder or abuse of the generative organs.

Linea nasalis (from upper border of ala nasi, curved to outer margin of the orbicularis oris) — strongly marked in phthisis and atrophy, inferior part indicates gastric disease, upper part affection of the upper part of intestine. Conjointly with retraction of the cheek, and with the L. oculo-zygomatica, the eyes being fixed and complexion wan, an indication of worms.

L. labialis (from angle of mouth to lower part of face) — in children, a thoracic affection with dyspnœa.

L. collateralis nasi (in a semicircular direction toward the chin, external to last two) — chronic and obstinate disease of thoracic or abdominal viscera.

Œdema of face and eyelids — albuminuria, sometimes in anæmia.

Transient redness or flushing of face — women suffering from menstrual irregularity, and at the critical period.

Hectic flush — phthisis, wasting, chronic affections.

Paleness of face — cold stage of fever, and acute inflammation, chronic diseases, especially Bright's disease, in convalescence.

Dingy white or greenish face — anæmia.

Yellow tint — jaundice.

Yellow at labial commissures and alæ nasi — slight hepatic derangement.

Citron tint — in cancerous affections.

Bluish hue — impeded venous circulation, as in asphyxia, Asiatic cholera, typhus fever, cyanosis.

Slate color — from the use of Nit. Silver.
Perpetual motion of eyelids — some cases of mania and idiocy.
Forcible closure of eyelids — photophobia.
Eyelids open — injury of portio-dura from paralysis of the orbicularis.
Paralysis of upper lip — lesion of third pair of nerves.
Epiphora (flowing of tears over the cheek) — obstruction of lachrymal duct, in initial stage of ophthalmia, and in some neuralgic affections of the eye, presence of a foreign body.
Nostrils dilating forcibly and rapidly — difficult respiration.
Itching nostrils — in children a sign of intestinal worms.

4. Region of the Throat.

Enlarged — some anginose affections, in the first months of pregnancy, and at the approach of puberty in females.
Violent pulsation of carotid arteries — acute mania, cerebral inflammation, hypertrophy of heart with dilatation of right ventricle, anæmia, sometimes in typhoid fever.
Pulsation of arteria innominata (above the sternum, in front and to the right of the trachea) — aortic regurgitation.
Circumscribed swellings — glandular enlargements.

5. Region of the Chest.

General expansion of one side — large pleuritic effusion.
Bulging at base of lung — gravitating pleuritic effusion.
Bulging at anterior superior parts of chest — emphysema.

Bulging in right hypochondrium — enlargement of liver.

Bulging in præcordial region — effusion into pericardium, or hypertrophy of heart.

Tumor about the junction of third rib, with right side of sternum — aneurism of ascending aorta.

Tumor between the base of scapula and the spine — aneurism of descending aorta.

Retraction of one side (usually the left) — after absorption of pleuritic effusion.

Depression or local retraction — absorption of circumscribed effusion, phthisis.

Respiration increased (healthy standard about 20 a minute) — dyspnœa, as in spasmodic asthma.

Respiration diminished — pleurisy and pleurodynia, paralysis of respiratory muscles, pneumonia, emphysema, pneumothorax, phthisis, etc.

Respiration jerking — spasmodic, asthma, obstruction of larynx and trachea, pleurodynia.

Respiration costal — abdominal inflammation and diaphragmatic pleurisy.

6. Abdominal Region.

General increase of volume — ascites, meteorism, tympanites (these latter known by resonance, and occur in adynamic diseases, peritonitis, intestinal obstruction, hysteria, etc.).

Enlargement of hypochondria — diseases of liver or spleen.

Enlargement in epigastrium — hysteria and cancer of stomach.

Enlargement in hypogastrium — distension of bladder, ovarian tumors, fecal accumulation, etc.

Diminished size — in most chronic diseases, as chronic dysentery, in lead colic (with hard muscles).

7. Genital Organs.

Enlarged penis in children — vesical calculus, masturbation.
Cartilaginous hardness of corpora cavernosa — from onanism.
Retraction of testicles — renal calculus.
Distended scrotum — hydrocele, hematocele, sarcocele.
Enlarged labia majora — general dropsy, local affections.

8. Extremities.

Immovable — paralysis.
Contracted and rigid — softening of brain, etc.
Œdematous — from embarrassed circulation.
Articulations swollen — rheumatism, hydrarthrosis, white swelling, etc.
Diminished in size — paralysis.

B. PHYSIOLOGICAL.

1. Functions of the Nervous System.

I. Sensation.

Morbidly augmented — acute inflammatory affections of brain and spinal marrow, idiopathic fevers, hysteria.
Tensile pain — phlegmonous inflammation.
Dull, heavy pain — enlarged viscera, internal tumor, effusion in serous cavities, in the loins previous to menstrual and hemorrhoidal discharges.
Smarting pain — skin deprived of cuticle, or under influence of irritants.
Lancinating pain — cancer and neuralgia.

SYMPTOMATOLOGY. 185

Boring pain — constitutional syphilis, rheumatism, gout, inflammation of periosteum, etc.
Contusive pain — from bruises, in acute diseases.
Itching and formication — cutaneous disorder.
Exaltation of vision — ophthalmia, inflammation of brain and meninges, some nervous affections.
Muscæ volitantes — affections of brain and optic nerve, dyspepsia.
Hearing painfully acute — cerebral inflammation, hysteria.
Hearing obtuse — in typhus fever.

II. Voluntary Motion.

Increase of strength — acute disorder, with delirium, cerebral inflammation, mania.
Debility — in most diseases.
Paralysis — indicative of lesion of brain or spinal marrow, as apoplexy, spinal softening, etc., or of injury to a nervous trunk, or it may be functional, as in some cases of hysteria.
Trembling — cold stage of fever, nervous affections, ataxic fevers, in old persons, action on the system of lead, mercury, strong coffee, alcoholic liquor, tobacco, and opium.
Rigidity of limbs — in upper extremities, a symptom of softening of the brain, cerebral extravasation, hysteria.

III. Reflex or Excito-Motory System.

Cramp — pregnant women, hysteria, painter's colic, etc.
Tetanus (another form of tonic spasm) — may be trismus when the muscles of mastication are affected, emprosthotonos, with the body bent forward, opisthotonos, if bent backward, and pleurosthotonos, with lateral curvature.

Clonic or temporary spasm — seen in convulsions of children, in hysteria, and some affections of the brain, in subsultus tendinum taking place in acute, ataxic diseases, in hiccough, etc.

Morbid rhythmical movements — disease of cerebellum or its commissures.

Reflected or sympathetic sensations, as pain at extremity of penis from calculus of bladder, pain in right shoulder from congested liver, in left shoulder from disordered stomach, etc., are numerous.

IV. Intellect.

Exaltation of affections — hypochondriasis.
Abolition of moral sensibility — mania, typhus fever.
Illusion and hallucination — insanity.
Exaltation of intellect — melancholia, sometimes at close of life.
Enfeebled intellect — typhus fever.
Delirium — diseases of brain and its meninges, typhus fever, the exanthemata; diseases of chest and abdomen.
Insomnia — mania, etc.
Drowsiness — typhoid fever, some affections of the brain, etc.

2. Function of Respiration.

I. Dyspnœa.

1. From the access of pure air impeded.

a. *Mechanical.*

Rigidity of parts — cartilages ossified, pleura indurated, rickety distortion.
Pressure of parts — tumors or dropsies of abdomen.

Obstructions of air-tubes — effusions, swellings, or tumors pressing on them, spasm of glottis or bronchi.

Compression of lungs — effusions or tumors in pleural sac, in pleurisy, hydrothorax, pneumothorax, aneurism, etc.

Alteration in tissue of lungs — enlargement of the vessels, effusions, as œdema, hepatization, tubercle, etc., altered structure, or emphysema, dilated bronchi, vomica, etc.

b. *Chemical.*

Deficiency of oxygen in the air — mephitic gas, rarefied air.

c. *Vital.*

Pain of parts moved in respiration — pleurodynia, pleuritis, peritonitis, etc.

Paralysis of muscles — injury of spinal marrow.

Weakness of muscles — prostration in ataxic fever, etc.

Spasm of muscles — tetanus, spasmodic asthma.

2. From the state of the blood.

a. *Mechanical.*

Obstruction to the passage of blood — diseases of heart and great vessels, tumors pressing on them.

b. *Chemical.*

An excessive venous state — violent exertion.

Deficiency of red particles — anæmia, chlorosis.

3. From the nervous relation of parts.

Excessive sensibility of par vagum — hysteric dyspnœa, cerebral fevers.

Defective sensibility of par vagum.—coma, narcotism, etc.

II. Cough.

Hollow or barking—in last stage of consumption, chronic bronchitis, in some nervous affections.
Sharp or ringing—in croup.
Hoarse—incipient catarrh, chronic laryngitis, anginous affections.
Wheezing—asthma.
Belching—some disease of larynx.
Paroxysmal—whooping-cough, hysteria.
On auscultation, bronchial (harsh, rapidly evolved, concentrated)—phthisis, pneumonia, pleurisy, dilatation of bronchi.
On auscultation, cavernous (hollow)—tubercular excavation, dilated bronchi.
On auscultation, amphoric (metallic or ringing)—broncho-pleural fistula, large, tubercular excavation.

III. Expectoration.

Scanty—first stage of acute affections of lungs.
Copious—decline of acute diseases of air-passages or lungs, in chronic affections.
Serous or watery—forming stage of bronchitis, pulmonary congestion, and vesicular emphysema.
Mucous—bronchitis and pneumonia.
Purulent—phthisis, third stage of pneumonia.
Nummular (like coin)—tubercular phthisis, bronchitis of measles, occasionally in chronic bronchitis.
Flocculent, muco-purulent—advanced phthisis.
Tubular—plastic bronchitis, pneumonia.
Whitish—beginning of acute affections of the lungs.
Yellowish or greenish—acute bronchitis.
Rusty—pneumonia.

Putrid smell — gangrene of lungs.
Faint and sweetish smell — bronchitis, and first stage of phthisis.
Alliaceous odor — broncho-pleural fistula.

IV. Pain.

Dull, heavy, aching, round the base of the chest — acute bronchitis.
Soreness in sternal region, and between the shoulders — acute bronchitis.
Sharp, lancinating, sudden, usually below the nipple — pleuritic.
Darting from anterior part of chest to interscapular region — in phthisis.
Constant pain between the shoulders — in phthisis, chlorosis, other chronic diseases.

V. Effects of Percussion.

Increased clearness of sound — in pneumothorax and emphysema.
Dulness of sound — pneumonia, pleurisy, phthisis, hydrothorax, etc.
Wooden sound — chronic pleurisy, with dense membranes.
Tympanitic — pneumothorax and emphysema.
Tubular — pleuritic effusion, tubercular excavation.
Amphoric — (imitated by filliping the inflated cheek) — tubercular cavities.
Cracked metal sound — cavities near the surface.

VI. Effects of Auscultation.

Exaggerated respiration — in portion of lungs adjoining those unfit for respiration.

Weak respiration — from obstruction to entrance of air to the part.

Suppressed respiration — when mucus clogs up a large bronchus.

Jerking respiration — incipient pleurisy, spasmodic asthma, tuberculous infiltration.

Incomplete respiration (inspiratory murmur deficient) — spasmodic asthma.

Bronchial respiration (like the top of sternum and root of lung naturally) — in pneumonia, tubercles, etc.

Cavernous and amphoric — tuberculous excavation.

Dry Rhonchi.

Sibilant — in bronchitis, from modified calibre of air-cells.

Sonorous — in bronchitis, from modified calibre of air-cells.

Dry, crackling (few in number, coexisting with inspiration) — first stage of phthisis.

Humid Rhonchi.

Crepitation (imitated by rubbing a lock of hair between the fingers near the ear) — pneumonia in stage of engorgement and of resolution.

Subcrepitant (more moist than last) — in capillary bronchitis, pneumonia at resolution, pulmonary apoplexy, œdema of lung.

Mucous — bubbling through liquid in bronchi of large size.

Cavernous or gurgling — same as last, but in the pulmonary excavation.

Friction-sound — from diseased pleura.

Diminished vocal resonance — in vesicular emphysema, pneumothorax.

Exaggerated vocal resonance, or bronchophony — tubercle, pneumonia in stage of hepatization.

Ægophony (nasal tone like a bleat of a goat) — pleuritic effusion.
Pectoriloquy (resonance, as in a hollow, and transmitted in articulate words) — tubercular caverns, and dilated bronchi.
Metallic, tinkling sound (imitated by striking gently a hollow glass vessel with a pin) — in pneumo-hydro-thorax, with bronchial fistula, in some excavations of the lungs.

3. Circulatory Functions.

I. Auscultation of the Heart.

a. *Its Impulse.*

(Is correspondent with the pulse at the wrist, unless mechanical impediments exist.)

Strong — in fevers and inflammations.
Small vibratile — after hemorrhage, in anæmia, etc.
Full, strong, heaving, and somewhat diffused — hypertrophy.
Still more powerful, felt over the whole præcordial region — hypertrophy with dilatation.
Feeble and diffused — ventricular dilatation.
Sharp, concentrated — in anæmic or nervous persons, atrophy of the muscular walls of the heart with fatty degeneration.
Visible at scrobiculus cordis — obstruction anterior to tricuspid valve.
Visible at scrobiculus cordis, as well as between the ribs of left side — disease of mitral valve.

b. *Its Rhythm.*

(The natural rhythm is a long sound, a short sound, and an interval.)

Altered rhythm — most frequent cause is valvular change, dilatation of heart and atrophy of walls of ventricles, effusions into the pericardium.

c. *Its Sound.*

(Natural sound represented by lub-tub lub-tub.)

Louder and clearer — dilatation of the cavities, with thinning of the walls, without valvular disease.

Clearer, but not louder — muscular atrophy of the parietes.

Decreased sounds — impeded action, hypertrophy.

(Unnatural sounds, or murmurs.)

Endocardial (or blowing murmurs, bellows murmur, rasping murmur, filing murmur, musical murmur), — indicative of valvular lesions, of diseased blood, as anæmia, of nervous disease of heart.

Exocardial (rubbing murmur and its varieties) — pericarditis, from the attrition of roughened surfaces.

II. Character of the Pulse.

a. *As to its Force and Intensity.*

Strong, resists compression by the finger. In inflammatory affections, especially of the parenchyma of the solid viscera, as lungs and liver, in the active hemorrhages. In plethoric and strong individuals, any derangement of circulation will cause it.

Weak (easily compressible) — disease with prostration, nervous and chronic affections, especially when caused by perverted nutrition, produced by fear, diseases of old men, women, and children.

Full, volume of artery seems increased — natural pulse of plethoric and tall persons, diseases with strong

SYMPTOMATOLOGY. 193

pulse, cerebral congestion and apoplexy, cardiac disease.

Small, opposite of full — often from narrowing of aortic orifice, in the serous phlegmasiæ, as peritonitis, pericarditis, inflammations of stomach, intestines, bladder, etc., in hysteria, hypochondriasis, and other nervous affections, in chlorosis, in cold stage of fevers, diseases with violent paroxysms of pain, a symptom of adynamic and ataxic diseases, and of purulent resorption.

Corded (hard, sharp, or contracted, giving a vibratory sensation to the fingers) — in the membranous phlegmasiæ, sanguine congestions, active hemorrhages, neurosis, lead colic, etc.

Soft (compressible or liquid, yields readily to pressure) — in adynamic affections.

b. *As to its Rhythm.*

Frequent — febrile and inflammatory disease, hemorrhages, etc.

Slow or infrequent — apoplexy, acute tubercular meningitis, some adynamic affections, sometimes in diseases of heart.

Unequal (dicrotous or double in beat) — convalescence.

4. Function of Digestion.

Tongue diminished in size (generally also trembling and dry) — typhus and other low fevers.

Tongue coated, etc. — Dr. Louis' observations indicate that the tongue does not show the true state of the stomach. This is a subject worthy of further investigation.

Appetite voracious — pregnancy, hysteria, and insanity.

Appetite diminished — most acute diseases.

Thirst increased — acute affections, especially of stomach and bowels — after hemorrhages, in diabetes.

Thirst abolished — some cerebral diseases with coma.
Vomiting — beginning of acute inflammatory and febrile affections, early pregnancy, in colic, cerebral diseases, hernia.
Pain aggravated by pressure — inflammation of viscera, peritonitis.
Pain relieved by pressure — overdistension, neuralgia, colic.
Tormina — acute colic.
Tenesmus — in dysentery.
Fæces watery — serous diarrhœa, Asiatic cholera.
Fæces mucous (like white of egg) — chronic inflammation of the colon.
Fæces hard and scybalous — constipation, colic, cancer of stomach, etc.
Fæces clay color — deficiency of bile.
Fæces yellow or dark brown — excess of bile.
Fæces dark green — from bile, after calomel in children.
Fæces red or streaked with blood — dysentery, when the blood is dark, and mixed with the fæces, it is usually from the upper part of intestinal canal.
Fæces pitchy black — melæna.
Fæces pure blood, unattended with colic — hemorrhoids.
Fæces semi-transparent and colorless, with whitish clots (like rice-water or turbid whey) — Asiatic cholera.
Fæces black — from iron as medicine.
Fæces with shreds of false membrane — dysentery and diarrhœa, biliary or intestinal calculi, worms, etc.
Fæces with fat — diabetes, phthisis.
Fæces fetid — adynamic diseases.

5. Urinary Secretion.

Suppression or diminution — most inflammatory and febrile diseases, dropsy.

Retention in the bladder — from paralysis, typhoid fever, hysteria, etc.
Increased amount — diabetes, cold stage of fevers, hysteria, from various passions of the mind.
Urine darker than usual in inflammatory affections; if much blood is abstracted during their progress, it becomes clearer; at the height of the inflammation it is clear and deeply colored; when it subsides, there is a yellow or reddish sediment of uric acid and urates.
Deposits of uric acid (red or yellow sand sediments) — fever, acute inflammation, rheumatism, phthisis, all the grades of dyspepsia, disease attended with arrest of perspiration, diseases of genital apparatus, from blows and strains of the loins, excessive indulgence in animal food, too little exercise.
Deposits of earthy phosphates (white sediment) — indicate a depressed state of the nervous energy of serious importance.
Deposits of oxalate of lime — digestive derangement.
Urine containing blood — hemorrhage of kidneys or urinary tract.
Albuminous urine — Bright's disease, dropsy after scarlatina, etc.
Mucous urine — irritated or inflamed state of genito-urinary mucous membrane.
Sugar in urine — dyspepsia; when excessive, diabetes mellitus.

6. Perspiration.

Profuse — acute rheumatism, decline of acute inflammations and fevers (the latter often critical).
Diminished — early stage of acute disease, dropsy, diabetes.
Night sweats — phthisis (profuse, debilitating).
Excessive acid odor — (rheumatism, gout).

Odor fetid — some adynamic fevers.
Odor mouldy — measles, scarlet fever.
Odor ammoniacal — sometimes in typhoid fever.
Odor peculiar in insanity.
Odor of chlorine or rottenstone — miliary.

7. Animal Heat.

General heat of surface — in fevers.
External local heat — in inflammation.
Forehead hot — cephalalgia.
Scalp hot — cerebral disease.
Integument of chest hot — thoracic inflammation.
Hands and feet hot — phthisis.
Peculiar acrid heat (burning the applied hand) — in typhus fever.
Chill — initial of fever, and of the phlegmasiæ, particularly pneumonia.
Temperature generally low — from languid circulation.
Coldness of hands and feet — in nervous and anæmic persons.

IX.

OUTLINES OF GENERAL PATHOLOGY AND THERAPEUTICS.

A. PRIMARY ELEMENTS OF DISEASE.

I. Properties of Contractile (Muscular) Fibre.

a. Irritability.

1. *Excessive.* Seen in excessive strength, as in delirium; or in quickness, as in convulsions or clonic spasm; or in unusual duration, in tonic spasm (cramp, catalepsy, and tetanus). *Remedies.* — If from flow of blood, antiphlogistic; firm pressure on muscles in cramp (masseter muscle in trismus). If from nervous irritation, narcotics and antispasmodics, especially stramonium, belladonna, sulphuric ether, and Indian hemp.
2. *Defective.* In force (weakness and paralysis) or in readiness to contract (as from opium, digitalis, etc., and some cerebral diseases). *Remedies.* — Repose, if from exhaustion; stimulants, as ammonia, brandy, etc. (often large quantities), electricity, cold water dash; strychnia and cantharides in paralysis (endermic application best).

b. Tonicity.

(Cold increases tonicity and impairs irritability.)

1. *Excessive. Remedies.* — Antimony, etc., to relax the fibre.

2. *Defective. Remedies.* — Tonics, especially cold, Peruvian bark, iron, the mineral acids, and generous living.

II. Properties of Nerves.

a. Sensibility (General).

1. *Excess.* Narcotics, mostly required, as opium, henbane, hemlock, etc. If vascular excitement, antiphlogistic treatment; with weakness, slow pulse, and absence of fever, tonics and stimulants as well as narcotics required. (Inhalation of ether the most powerful anodyne.)
2. *Defective* (as in coma, etc.). Sometimes depletion; if no disease, mental excitement, bodily exertion; the cold dash, and friction; from narcotics and retained excrement; use purgatives, diuretics, emetics, etc. If anæmia, stimulants.
3. *Perverted Sensibility* (illusory or depraved sensations). Chalybeates, etc.; narcotics, etc., as palliatives.

b. Local Sensibility.

1. *Excessive* (from disease, etc., as the pain of pleurisy, etc.). If from inflammation, antiphlogistics; if remaining after inflammation, anodynes (endermic application of morphia often useful: remove the cuticle by a blister, and apply one or two grains of a soluble salt of morphia; as the acetate or hydrochlorate; repeat once or twice daily, and keep the surface moist); counter-irritation and warmth (as gastrodynia relieved by a sinapism at the pit of the stomach). In a weak circulation, especially if the pain be intermittent, tonics are useful (as neuralgia treated with quinine, and iron; hemicrania with quinine, or liquor arsenicalis).

c. Voluntary Motion.

1. *Excessive.* Depletion, antimonials, cold to the head, if determination of blood. If more nervous, narcotics.
2. *Defective.* Excite the nervous centres through the circulation, as by stimulants, etc. (Hysterical coma often removed by turpentine injection, or croton oil purge, which acts both as a revulsive to the vessels and a stimulant to the nerves.)
3. *Perverted Volition.* Treatment various; in delirium tremens by narcotics, as opium; in chorea, by nervous tonics, especially iron and zinc.

d. Reflex Action.

Connected with organic life. The contractions of all the sphincters, and the regular action of the muscles of respiration depend on it. (A nervous influence, independent of the will conveyed by afferent nerves from the surface to the spinal marrow, and reflected from it through the afferent nerves to the muscles of the parts.)

1. *Excess.* Seen in spasm of throat in hydrophobia, tetanus, hysteria, etc.; in convulsive motions of lower limbs when tickled, etc., in paraplegia, etc. Also in epileptic and apoplectic convulsions, which are centric when resulting from diseases in the head or loss of blood; or eccentric when from irritation of the extremities of afferent nerves; as from teething, intestinal, uterine, and renal irritation, passing a bougie, sometimes, etc.; also in partial spasms and sympathetic irritation of distant parts.

If these inordinate reflex actions are general or extensive, as convulsions, tetanus, and paraplegia, we refer them to undue excitement, or erethism of the spinal marrow; the more partial examples (sympathetic irritation, etc.) may arise from a small portion of it only, or of the afferent or efferent nerve of the part. Increased flow of blood to the medulla, or its nerves, or the branches of the sympathetic nerve; the direct action of poisons, as strychnia; mechanical irritation on the spinal marrow, or its nerves (as in tetanus, tumors, and spicula of bone in spinal canal, etc.), may cause this excitement. The involuntary excito-motory property is also accumulated by rest and sleep. Hence narcotism, injury of the spine, sedentary habits, too much sleep, etc., by suspending volition, may cause a morbid excess of involuntary nervous power, and develop convulsive and spasmodic symptoms, which are the result of its overflow.

Remedies. — Often antiphlogistic, because often dependent on determination of blood. If more purely nervous, as tetanus, etc., a narcotic used (hydrocyanic acid, woorara, resin of Indian hemp, conium, etc., reduce the power of the spinal system, and cause general relaxation of muscles, but they may destroy life by arresting respiration; useful, however, in small doses in slighter irritations, as vomiting, nervous palpitation, and hiccough). Extract of belladonna and stramonium, useful in convulsive cough and spasmodic asthma, and combined with opium in the spasms of colic, dysentery, and dysuria.

In weak subjects, without inflammation, medicines which act as stimulants to the heart, and vessels, and cerebral functions, and also as sedatives to the medullary system (the stimulant antispasmodics, as ether, ammonia, musk, essential oils, external heat and counter-irritation, etc.). Tonics also reduce the excitability of the spinal excito-motory system, especially metallic tonics, as iron, nitrate of silver, sulphate and oxide of zinc, and sulphate of copper, cold baths, change of air, and exercise useful.

2. *Defective reflex actions;* seen in paralysis of sphincters, eyelids, and muscles of respiration; and extreme debility from fatigue, excitement, or directly depressing influences.

Remedies. — Stimulants, narcotics, tonics (indiscriminate use of narcotics, hazardous in extreme weakness); should be preceded or combined with stimulants; those least depressing to be preferred, as opium; give suitable nourishment also in liquid form.

e. Reflected (or Sympathetic) Sensation.

Reflex action referred to motion, but the impressions which cause sensation may be reflected in a similar manner: thus, ascarides in the rectum cause itching of the anus, congestion of the liver, often a pain in the right shoulder-blade, and the pains of angina and gastrodynia often extend to the whole chest; the former especially radiates to the left arm.

Remedies. — 1. Those that remove the irritating cause. 2. Anodynes, by deadening sensibility (the efficacy of trisnitrate of bismuth and hydrocyanic acid in gastrodynia, and some kinds of angina, not referable to a narcotic property). Tonics are often useful, as morbid sympathies, like other nervous disorders, are exalted by weakness or irregularity of the circulation.

III. Properties of Secretion.

1. *Excessive Secretion* weakens, from the drain it causes from the blood. Its effects may be forward on the parts to which the secretion goes, or backward on the organ which secretes it and the blood from which it is formed.

(Forward effects of excessive secretion of bile seen

in bilious diarrhœa or cholera; of profuse mucous secretion in the intestines in simple diarrhœa; in the bronchi in dyspnœa and cough; in the stomach, seen in pyrosis or water-brash, etc. Backward effects seen in torpid bowels after diarrhœa.)

Remedies. — If dependent on the quantity and quality of the blood, depletion, derivation, and evacuants. (The excessive secretion then a means of relief, and arrested by increasing it, as a purge of calomel will stop a bilious diarrhœa, from an engorged liver.) If from nervous or other irritation, causing weakness and disturbance of the functions, it may be checked by tonics and astringents (as cold to the part, alum, superacetate of lead, sulphates of zinc and copper, gallic acid and tannin, vegetables which contain tannin, etc., mineral acids, etc. These act by direct application, as in diarrhœa or leucorrhœa, or through the circulation. Some agents, without a general astringent effect, diminish the secretion of particular organs, as opium, which remarkably lessens the secretion of the liver, and sometimes that of the kidneys).

If excessive secretion have caused febrile disturbance, means to increase other secretions may restore a proper balance. Thus, in bilious cholera, saline diuretics and diaphoretics are serviceable; in renal irritation, with copious secretion of lithic acid, blue pill to augment the bile is often beneficial (combinations of medicines more useful in any disturbance of secretion, especially if long continued, as mercurials with diuretics, antimonials with salines, etc.).

2. *Defective.* May cause general plethora, or local congestions, leading to dropsical effusions, fluxes, hemorrhages, or inflammations. Forward effects instanced in disorder in the latter stages of digestion, from deficiency of bile. Backward effects often seen in con-

gestion of the organ; and very remarkably in the case of the excretions, as urine and bile, which poison the blood when retained, causing typhoid symptoms, extreme depression, coma, and death. If the suppression be incomplete in the latter instances, the poisoning process is more tardy, producing various functional and visceral derangements, as delirium, or lethargy, dyspnœa, palpitation, vomiting, diarrhœa, etc. (The excrementitious matters may then be detected in the blood, and other parts of the body; as the color of bile in the textures in jaundice; urea in the blood, etc., in glandular degeneration of the kidneys, etc. Gout, rheumatism, degeneration, dropsies, etc., are often caused by various degrees of defective excretion.)

Remedies. — If from defective supply of blood, stimulants, etc.; if from inflammation or congestion, depletion or derivation. Often the first disorder is in the secreting structure itself, and the remedy must be those agents which increase the respective secretions; as mercury for the liver; colchicum, nitre, etc., for the kidneys; croton oil, jalap, sulphate of magnesia, etc., for the intestines, etc. (These specific stimuli in excess, or too long continued, may cause not only general weakness, but also an exhaustion of the vital properties which they excite; as long or excessive use of mercury causes torpidity of the liver; of purgatives, imperfect action of the bowels; of diuretics, scanty, albuminous, or watery urine;—hence they should be intermitted and alternated with tonics, as bitters with mercurials; chalybeates with saline aperients and diuretics.) In chronic cases, medicines which are inferior in efficacy to be preferred, because less exhausting — (examples: taraxacum, iodine, sarsaparilla, nitric and nitro-muriatic acids). Where defective secretions are not readily restored, they may sometimes be compensated for by artificial substitutes. Thus ox-gall, aloes, and soap, or toasted bacon at breakfast, promotes the action of the intestines, in defective secretion of bile; and defective secretion of mucus may be remedied by mucilage, etc.

3. *Perverted Secretion* often accompanies excess and defect. In febrile diseases, the secretions of the kidneys and the alimentary canal are altered as well as diminished; inflammation and determination of blood change as well as increase the secretion from mucous membranes, rendering it more saline and sometimes albuminous, etc. Altered secretion may be unfit for use; as a thin, acrid mucus irritates instead of protecting the membrane, as in coryza and mucous diarrhœa; viscid, dry mucus obstructs the tubes; altered gastric juice causes indigestion; sebaceous matter accumulating in the follicles of the skin causes irritation, inflammation, etc.

Remedies. — Usually those which increase secretion. In some cases tonics may be advantageously combined with them. Such a combination is presented in most of those remedies called alteratives.

IV. Constituents of the Blood.

The principal constituents of the blood, necessary to be mentioned here, are the red particles, fibrin, and albumen, either in excess, defect, or alteration. The other constituents of the blood are oil, salts, and water. In malignant cholera, the defect of saline matter and water, owing to the excessive evacuations, seems to be the cause of the obstructed circulation, lividity, and collapse, and hence the temporary efficacy of injection of saline solutions into the veins of such patients.

a. Red Particles.

The red particles of the blood are distinct structures — living cells, floating in the liquor sanguinis; they have a tendency to cohere in piles or roleaux in fresh-drawn blood, and this tendency is strongest in blood taken from a person affected with inflammation.

1. *Excess.* Seen in sanguineous plethora. A slight increase has been detected in the early stage of inflammations and fevers, especially eruptive fevers, as measles and scarlatina.

    Remedies. — Bloodletting the speediest agent. Low or vegetable diet and the antiphlogistic regimen generally: saline medicines, much diluted and taken copiously, have a remarkable effect.

2. *Defect.* Seen in the lymphatic temperament, also after a great loss of blood, in chlorosis and other anæmic states, in scrofulous and tubercular diseases, in the latter periods of fevers, and after severe inflammations, in granular degeneration of the kidney, etc. Known by paleness of parts naturally red, pallid or sallow complexion, a weak state of the functions generally.

    Remedies. — Air; light nourishing food, especially brown meats; tonics, particularly iron (Quevenne's metallic iron, *Ferrum per hydrogen*).

3. *Alteration.* Of the red particles is evinced by change of color in the blood, and change of form in the individual corpuscles, as seen by the microscope.

    Seen in scurvy; in the Walcheren and other malignant fevers, in cachæmia, from malarious influence, generally in connection with a diseased spleen; in congestive typhoid fevers, etc.

    Remedies. — Saline medicines have been recommended in typhoid and malignant fevers; but remedies to increase the excretions in connection with those mentioned under the last head, are less questionable; in malarious and anæmic cachæmia; the use of purgatives and diuretics, combined with chalybeate tonics, has produced the best effects.

b. Fibrin.

But little difference between fibrin and albumen, in chemical composition, yet fibrin is distinguished by its being organizable or susceptible of life. It causes the coagulation of the blood; it constitutes the buffy coat and coagulable lymph; and is probably the material by which chiefly the textures are nourished and repaired.

1. *Excess.* In all true inflammatory diseases, especially those of a sthenic character, and in acute rheumatism. There is a relative excess also in diseases connected with a deficiency of red particles.

Remedies. — Bloodletting and low diet; yet fibrin is less reduced by them than excess of red particles. Remedies which increase the more solid secretions probably diminish the fibrin.

2. *Defect.* Seen in fluidity, or but slight coagulation of blood when drawn, or in asthenic tendency to hemorrhage, and unmanageable oozing of blood from an accidental wound, etc., in cases of poisoning with hydrocyanic acid, etc., in adynamic fevers, etc., in cases of asphyxia, cyanosis, etc.

Remedies. — Assist the functions on which the supply of fibrin depends. If the digestive organs will bear them, meat, eggs, bread and other articles abounding in protein; assist digestion and assimilation by stimulants, bitters, quinine, and the mineral acids; help respiration by the access of pure, cool air; avoid fatigue; secure sleep, if necessary, by

narcotics; toxicological means, of course, if called for.
3. *Alterations.* Seen in the varieties presented by the buffy coat, and contractions of the clot of blood. Also in the varieties of the reparative process. False membrane, deposits, etc., in a healthy subject, may be euplastic, or in a high degree organized and healthy. But in many instances the nutritive material is caco-plastic, or susceptible of only a low degree of organization, as in induration, from chronic inflammation, in fibro-cartilage, cirrhosis, gray tubercle, etc. It may also be aplastic, or not organizable at all, as in pus, curdy matter, yellow tubercle, etc.

Remedies. — The increased properties of separation and contraction manifested by blood in inflammation, are reduced by bloodletting, etc. Yet if antiphlogistic-remedies do not remove local inflammation, they may render its product more injurious by lowering its plasticity. Hence the necessity of endeavoring to remove inflammations before they become chronic, and when there is risk of such event, improving the condition of the blood by a tonic and nutritive plan, conjoined with local antiphlogistic measures. A similar tonic treatment is indicated in scrofulous, chlorotic, and other cachectic states, where the fibrin is relatively copious with a tendency to aplastic deposits. In addition, remedies likely to keep the fibrine dissolved, as alkalies, and iodide of potassium, are advisable, although the efficacy of these means has not been fully proved.

c. Albumen.

1. *Excess.* Exists in most cases of inflammations, and fevers, especially during their more active stages. Its increase is not, however, in proportion to that of fibrin. Its excess in cholera is due to the removal of

the water of the blood. Very poor living, extensive hemorrhage, and other drains will reduce it.

2. *Defect.* Met with in cases of albuminuria, and in diabetes: it seems to be a chief constituent of the dropsical diathesis.

Remedies. — Those which restrain wasting discharges and improve general nutrition. Cod-liver oil recommended for the last purpose.

V. Changes in the Blood by Respiration.

The change of venous into arterial blood is never in excess, for the activity of the respiration is adapted to the rapidity of the circulation, and the corresponding need of change.

Defect of the change is the essence of asphyxia or apnœa.

Remedies. — Rest, fresh air and sedative medicines (as digitalis, hyoscyamus, etc.), or antispasmodics. Sometimes an enfeebled circulation may require stimulants, or an engorged venous system calls for depletion. In suspended animation from drowning, etc., artificial respiration, frictions, the warm bath, stimulants, etc.

VI. Changes in the Blood by Excretion.

See Sect. III. Property of Secretion.

Other changes may be produced in the blood from the transformation of the chyle and of the textures, including the processes of nutrition and reparation (probably the cause of gout, diabetes, and obesity). The presence of foreign matters in the blood, also, may excite various contagious, epidemic or endemic diseases; but too little is yet ascertained to supply any certain knowledge.

b. Proximate or Secondary Elements of Disease.

I. Anæmia.

The exciting causes of anæmia are circumstances which injure or withdraw the blood; profuse discharges of other fluids; scanty or poor food; impure air; chronic diseases, and uterine irregularity, as chlorosis.

The general symptoms are weakness, both muscular and organic; defective nutrition; and imperfect sanguification; the nervous system is also frequently excited.

Remedies. — (See A, Section IV.) Those which increase the constituents of the blood.

II. Hyperæmia or Excess of Blood.

This may be general (plethora), with increased motion (sthenic), or with diminished motion (asthenic); or it may be local, with diminished motion (congestion), or with increased motion (determination of blood). The results of these may be hemorrhage, flux, dropsy, etc. Another variety of hyperæmia may be distinguished by an altered or perverted action of the vessels. This is chiefly local, and includes inflammation.

Remedies. — For plethora, bloodletting and other evacuants. In the sthenic kind, sedative and relaxing remedies are also indicated, but in the asthenic, tonics, and even stimulants; or alterative aperients, as mild mercurials, with rhubarb, aloes, or senna, salines, and taraxacum, iodide of potassium, etc.; may prepare the way for various tonics.

For congestion, the most important means are those which contribute to a removal of its cause, as the loosening of a ligature, reduction of a compressing tumor, moderating the action of a diseased heart, or restoring the secretion of the liver, etc. In congestion from atony of the vessels, a change of posture sometimes gives relief, as in congestive fevers, when the head is affected, it should be elevated; congested uterine, or hemorrhoidal, vessels, and varicose limbs, are assisted by the recumbent posture. Pressure, as by bandages, etc., is also useful at times. Astringents are sometimes useful by increasing the contractility of the vessels. Stimulants also are often very effective, as diluted spirit lotion to a congested conjunctiva, capsicum gargle to a congested throat, or a stimulant wash to a purple sore, etc.

For determination of blood, the removal of stimuli or irritants from the part, or the reduction of their action by soothing or diluent remedies, is the first indication. The atonic distension of the arteries supplying the part may be relieved by cold, astringents, and derivants; as cold lotions to the head, and the hot footbath in determination to the head. Evacuants, also, and frequently bloodletting, are indicated as derivants.

Inflammations. — In incipient inflammation, for the congestion, astringents, stimulants, or evacuants may be useful. For the irritation of the nerves and vessels, sedatives, derivatives, and evacuants.

In local inflammation, the remedies for congestion and determination are applicable For impeded circulation in a part, moist heat, and other stimulants; but, for increased circulation, the remedies for determination. In inflammation with fever, general bloodletting and other evacuants are called for; relaxants, as antimony, etc.; low diet, etc.

(A remarkable fact has been discovered by Dr. Marshall Hall, viz., that in inflammatory diseases a much larger quantity of blood may be drawn without producing syncope, than can be taken in health, or in other diseases.) He says: "In cases in which it is doubtful whether the pain or other local affection be the effect of inflammation or of irritation, the question is immediately determined by placing the patient upright, and looking upward, and bleeding to incipient syncope. In inflammation much blood flows; in irritation very little." This he considers a rule for bloodletting, a guard against undue and inefficient bloodletting, and a "source of diagnosis, in the fullest sense of the word."

The following table shows the results of his investigations, as to the tolerance of bloodletting in different diseases, before incipient syncope.

I. Augmented Tolerance.

1. Congestion of the brain, ℥xl–l.
2. Inflammation of serous, synovial, and fibrous membranes, ℥xxx–xl.
3. Inflammation of the parenchyma of organs (brain, lung, liver, mamma, etc.), ℥xxx.
4. Inflammation of the skin and mucous membranes (erysipelas, bronchitis, dysentery), ℥xvj.

II. Healthy Tolerance.

This depends on the age, sex, strength, etc., and on the thickness of the parietes of the heart; and is about ℥xv.

III. Diminished Tolerance.

1. Fevers and eruptive fevers, ℥xj–xiv.
2. Delirium tremens and puerperal delirium, ℥x–xij.
3. Laceration or concussion of the brain. Accidents, before the establishment of inflammation. Intestinal irritation, ℥viij–x.

4. Dyspepsia, chlorosis, ℥viij.
5. Cholera, ℥vj.

The exhaustion from long-continued inflammation often renders stiumulants and tonics necessary; as also the depression arising from the influence of purulent or gangrenous matter. The effused products of inflammation require evacuants, attenuants, alteratives, stimulants, friction, etc.

Varieties of *Inflammation.* — The sthenic form requires of course all the antiphlogistic measures, but in the asthenic form, local bloodletting is better than general, which is illy borne. Antimony or mercury, and blisters, form the chief treatment. The diet, though light, should not be too spare.

Erysipelatous inflammation is generally asthenic, and often requires stimulant and tonic agents. The local treatment consists of punctures and incisions; cauterization by nitrate of silver; and mercurial ointment, which is supposed to modify the character of the poison. A solution of sulphate of iron in water, one ounce to a pint of water, applied to the part by moistened rags, acts like a charm.

The aphthous inflammation of children is to be treated by aperients, with a local application of borax, or a weak solution of sulphate of zinc.

Scrofulous inflammation owes its peculiarity to a degraded condition of the plasma, or nutritive material of the blood; and hence it is most benefited by tonics, nourishing diet, etc. Cod-liver oil, iodide of potassium, etc., are medicines in repute.

Rheumatic and gouty inflammation require means to eliminate the morbid matter from the system, as mercury and colchicum. Active antiphlogistic measures are often necessary, however, before these medicines will act.

Gonorrhœal inflammation requires mild antiphlogistic

and demulcent measures at first, and astringent injections and terebinthinate remedies afterward. Mercury is the chief specific in syphilis.

c. Structural Disease.

Including increased nutrition (hypertrophy), diminished nutrition (atrophy), and perverted nutrition, would open a field too extensive for a table like the present. It pertains, too, rather to the department of morbid anatomy than to pathology proper. (See A, Section IV.)

Modes of Death.

I. Beginning at the heart. 1. Suddenly (syncope). This is instantaneous; the subject suddenly turning pale, falling back, or dropping down, and expiring with one gasp.

2. Gradually (asthenia). The symptoms are, — increasing weakness of body and mind, with perhaps no marked derangement in any particular function; increased frequency, and diminishing strength of the pulse; the face, lips, etc., becoming paler and paler, or of a peculiar sallowness; extremities become cold and œdematous; tongue often dry and brown, or furred, and the mouth aphthous; excretions imperfectly voided at first, then the sphincters lose their power, and the discharges are involuntary; general sinking.

II. Beginning at the breathing apparatus (asphyxia or apnœa). Symptoms: increased feeling of suffocation; face, neck, etc., congested, and changing from red to purple, and from purple to livid; stupor; reduction of temperature; weak and irregular pulse; rapid reduction of muscular strength.

III. Beginning at the brain (coma). Its symptoms are those of interrupted function of the brain, insensi--

bility, and suspension of voluntary motion, the heart's action not being materially impaired. The excito-motory system of the medulla is often affected, as well as the sensorial and voluntary functions; hence respiration is interrupted, convulsions sometimes ensue, and the sphincters are relaxed.

IV. Beginning at the medulla (paralysis). This mode, like that of the last, is really by apnœa, but the excito-motory function is the first to fail. Of course there can be no respiration when this ceases.

V. Beginning with the blood (necræmia). The symptoms are typhoid, putrid, or malignant; a congested surface, the color being dusky or livid; exanthematous patches on the skin, or petechiæ; echymoses, or oozing of thin, bloody fluid from the gums, nostrils, etc.; extreme prostration; obtuse senses and mental faculties; sometimes with delirium and twitching of the limbs; half-closed eyes and dilated pupils; frequent and unequal respiration; no appetite; intense thirst; a dry, brown tongue with dark sordes on the lips and teeth; progressive fall of temperature; cold, clammy, and fetid perspiration; hiccough; subsultus tendinum; scanty, offensive urine; involuntary discharges.

INDEX.

ABBREVIATIONS, 10
Abdominal symptoms, 183
Abscess, 16
Absinthe, 30
Absinthium, 30
Acacia, 30, 148
Accumulative effects, 9
Acer Pennsylvanicum, 31
Acetate of copper, 71
    lead, 110
    morphia, 103
    potassa, 112
    soda, 125
Acetic acid, 33
    ether, 31
Acetum, 31
    aromaticum, 31
    cantharidis, 31
    colchici, 31, 161
    destillatum, 32
    lobeliæ, 32
    opii, 32, 161, 162
    sanguinariæ, 32
    scillæ, 32, 161
Achillea millefolium, 33
Acids, 12, 139
Acidum aceticum, 33, 147
    aceticum aromaticum, 33
    aceticum camphoratum, 33
    aceticum dilutum, 33
    arseniosum, 33, 146
    benzoicum, 34
    carbazoticum, 34
    carbolicum, 34
    carbonicum, 139

Acidum chromicum, 34
    citricum, 34
    gallicum, 35
    hydriodicum dilutum, 35
    hydrochloricum, 35. 139
    hydrochloricum dilutum, 35
    hydrocyanicum dilutum, 35
    lacticum, 36
    muriaticum, 35, 139
    nitricum, 36, 139
    nitricum dilutum, 37, 139
    nitro-muriaticum, 37, 139
    phosphoricum, 139
    phosphoricum dilutum, 139
    sulphuricum, 37, 139
    sulphuricum aromaticum, 37, 139
    sulphuricum dilutum, 38, 139
    sulphurosum, 38
    tannicum, 38
    tartaricum, 39
    valerianicum, 39
Aconite, 39
Aconiti, 39
Aconitia, 39
Aconitum napellus, 39
Actæa, 39
Adansonia digitata, 39
Adeps, 39
Adiantum pedatum, 40
Ægle marmelos, 54
Æsculus hippocast., 40
Æther sulphuricus, 40
Æthers, 148
Agaric, 40

216　INDEX.

Agathotes chirayta, 64
Agave Americana, 40
Age, in prescriptions, 5
Agrimonia eupatoria, 40
Agrimony, 40
Ailanthus glandulosa, 40
Ajuga chamæpitys, 40
Akazga, 41
Albumen, 207
Alchemilla vulgaris, 41
Alcohol, 14, 41, 148
　amylicum, 41
　dilutum, 41
　methylic, 41
Alcoholism, 20
Alcornoque, 42
Aletris, 42
Aleurites triloba, 42
Alisma plantago, 42
Alkalies, 12, 140
Alkaloids, vegetable, 154
Alkanet, 42
Alliaria officinalis, 42
Allium, 42, 132
Almonds, 46
Alnus rubra, 42
Aloe, 43
Alteration of fibrin, 207
　red particles, 205
Althææ, 43
Alum, 43, 142
　curd, 43
　root, 86
Alumen, 43, 142
　exsiccatum, 43, 142
Alumina, 142
Aluminæ sulphas, 44
Amaranthus hypochondriacus, 44
Amber, 130
Ambergris, 44
Ambrosia trifida, 44
American aloe, 40
　centaury, 120
　columbo, 80
　dittany, 71
　hellebore, 136
　holly, 91

American ipecacuanha, 83
　poplar, 112
Ammoniacum, 44, 141
Ammoniæ arsenias, 44
　benzoas, 44
　bicarbonas, 44
　boras, 44
　carbonas, 45, 141
　hydrochloras, 45, 141
　murias, 45, 141
　phosphas, 45
　valerianas, 45
Ammoniated copper, 72
　iron, 85
Ammonii bromidum, 46
　iodidum, 46
Ammonio-ferric alum, 77
Ampelopsis quinquefolia, 46
Amygdalæ, 46
Amylum, 46, 147
Anacardium occidentale, 46
Anæmia, 17, 209
Anagallis arvensis, 46
Anchusa officinalis, 46
　tinctoria, 42
Andira inermis, 56
Andromeda arborea, 46
Anemone pratensis, 46
Anethum, 46
Angelica, 47, 49
Angustura, 47
Animal charcoal, 59
　heat, 196
Aniseed, 47
Anisum, 47
Annotta, 47
Antennaria margaritacea, 47
Anthemis, 47
Antimonii et potassæ tar., 47, 146
　præcipitatum, 146
　sulphuretum, 146
Antimonium sulphuratum, 48
Antimony, 13, 146
Antirrhinum linaria, 48
Aphasia, 21
Apocynum androsæmifolium, 48
　cannabinum, 48
Aqua acidi carbonici, 155

INDEX. 217

Aqua ammoniæ, 49
    amygdalæ amaræ, 155
    camphora, 155
    cinnamomi, 155
    fœniculi, 155
    menthæ pip., 155
    menthæ virid., 155
Aquæ, 49
    rosæ, 155
Aquilegia vulgaris, 49
Aralia nudicaulis, 49
    spinosa, 49
Arctium lappa, 49
Areca nut, 49
Argemone Mexicana, 49
Argenti cyanidum, 50
    iodidum, 50
    nitras, 50, 145
    nitras fusus, 50, 145
Argenti oxidum, 51, 145
Armoracia, 51
Arnica, 51
Aromatic confection, 168
    sulphuric acid, 37
    vinegar, 33
Arrow poison, 72
Arrowroot, 100, 147
Arseniate of ammonia, 44
Arsenic, 13, 146
Arsenici iodidum, 146
Arsenious acid, 33, 146
Arum, 51
Asarabacca, 51
Asarum canadense, 52
Asclepias, 52
    curassavica, 52
    gigantea, 57
Ash, 80
Asparagus, 52
Aspect of patient, 179
Assacon, 87
Assafœtida, 52
Aster puniceus, 53
Astringents, 153
Atomizers, 53
Atropia, 53
Atropiæ sulphas, 53
Attitude of patient, 179

Aurantii cortex, 53
Aurum, 83
Auscultation of chest, 189
    of heart, 191
Australian gum, 76
Avenæ farina, 53
Azedarach, 53

BAEL FRUIT, 54
    Balm, 101
Balmony, 63
Balsam apple, 102
Balsams, 151
Balsamum Peruvianum, 53
    tolutanum, 54
Baneberry, 39
Baobab, 39
Baptisia tinctoria, 54
Barberry, 55
Barley, 87
Barosma, 56
Baryta, 12, 142
Bastard ipecacuanha, 52
Bay-berry, 104
Bay rum, 169
Beaked hazel, 69
Bear's foot, 85
Bear's whortleberry, 134
Bebeeru, 54, 105
Beech drops, 107
Belæ fructus, 54
Belladonna, 54
Benne, 125
Benzin, 55
Benzoate of ammonia, 44
Benzoic acid, 34
Benzoin, 55
Benzoinum, 55
    odoriferum, 54
Benzole, 55
Berberis vulgaris, 55
Betel nut, 49
Bethroot, 133
Betula, 55
Biborate of ammonia, 44
Bibron's antidote, 15
Bicarbonate of ammonia, 44
    potassa, 112

Bicarbonate of soda, 126
Bichloride of platinum, 110
Bichromate of potassa, 112
Birch, 55
Bird manure, 84
Bisenna, 102
Bismuth, 13, 28, 146
Bismuthi subcarbonas, 55
    subnitras, 55, 146
    valerianas, 55
Bistort, 55
Bisulphate of potassa, 55
    quiniæ. 117
Bisulphuret of carbon, 55
Bitartrate of potash, 112
Bitter almond water, 155
    ash, 56
    candy-tuft, 91
    cucumber, 67
    polygala, 111
Bittera febrifuga, 56
Bittersweet, 74
Black alder, 116
    berry, 119
    birch, 55
    haw, 136
    hellebore, 85
    oxide of mercury, 89
    pepper, 109
    snakeroot, 64
Bladder diseases, 16
Bladder-wrack, 80
Blessed thistle, 62
Blistering fly, 59
    plaster, 175
Blood diseases, 17
    root, 121
Blue cohosh, 62, 96
    flag, 94
    gentian, 82
Bole armenian, 56
Boletus, 40
Bone diseases, 18
Boneset, 76
Borage, 56
Borago officinalis, 56
Borate of soda, 126
Borax, 126

Boundou, 41
Box, 56
Brain diseases, 19
Brandy, 128
British oil, 105
Bromide of ammonium, 46
    potassium, 115
Bromine, 56, 143
Brominium, 56
Broom tops, 123
Brown mixture, 164
Bryony, 56
Buchu, 56
Buck bean, 102
    thorn, 118
Buglass, 46
Bugle weed, 98
Burdock, 49, 96
Burgundy pitch, 110
Burning bush, 76
Burnt sponge, 128
Butterfly weed, 52
Butternut, 95
Button bush, 62
    snakeroot, 75, 97
Buxus sempervirens, 56

CABBAGE-TREE BARK, 56
    Cactus grandiflora, 56
Cadmii iodidum, 56
    sulphas, 57
Caffea, 57
Cahinca, 57
Calabar bean, 109
Calamina, 57
    præparata, 57, 125
Calamine, 57
Calamus aromaticus, 57
Calcis carbonas præcipitata, 57
    phosphas præcipitata, 57, 142
Calendula officinalis, 57
Callitriche verna, 57
Calomel, 88
Calomelas, 88
Calotropis gigantea, 57
Calumba, 57
Calx chlorinata, 58, 142
Camphor, 58

INDEX. 219

Camphora, 58
Cancer root, 107
Canella, 58
Canna, 147
Cantharis, 59
    vittata, 59
Capsicum, 59
Caraway, 60
Carbo animalis, 59, 147
    ligni, 59, 147
Carbolic acid, 18, 21, 24, 34
Carbonate of lead, 111
    lime, 57
    lithia, 97
    magnesia, 98
    potassa, 113
    soda, 126
    zinc, 57
Carbonic acid water, 155
    oxide, 14
Cardamine pratensis, 60
Cardamom seed, 60
Cardamomum, 60
Carota, 60
Carrageen, 64
Carrot, 60
Carthamus, 60
Carum, 60
Carya, 60
Caryophyllus, 60
Cascarilla, 61
Cassia, 61, 65
    fistula, 61
Castanea, 61
Castor, 61
    oil, 150
Castoreum, 61
Catalpa cordifolia, 61
Cataria, 61
Catawba tree, 61
Catechu, 61
Catnip, 61
Caulophyllum thalictroides, 62
Ceanothus Americanus, 62
Cedron, 62
Celandine, 63, 123
Celastrus scandens, 62
Centaurea benedicta, 62

Centaury, 62
Cephalanthus occidentalis, 62
Cera flava et cera alba, 62
Cerata, 63
Cerate of carbonate of zinc, 176
    subacetate of lead, 177
Cerates and ointments, 175
Ceratum adipis, 175
    calaminæ, 176
    cantharidis, 175
    cetacei, 175
    plumbi subacetatis, 177
    resinæ compositum, 175
    sabinæ, 175
    saponis, 177
    zinci carbonatis, 176
Cerii oxalas, 63
Cetaceum, 63
Cetraria, 63
Chalk mixture, 163
Chamomile, 47
Changes in blood by excretion, 208
    respiration, 208
Character of the pulse, 192
Charcoal, 59, 147
Chelidonium majus, 63
Chelone glabra, 63
Chenopodium, 63
Chestnut leaves, 61
Chest symptoms, 182
Chicory, 64
Chimaphila, 64
Chinquapin, 61
Chiococca racemosa, 57
Chiretta, 64
Chloral, 64
    hydrate, 64
Chlorate of potassa, 113
Chloride of lime, 58
    mercury, 88
    sodium, 127
    zinc, 137
Chlorinated lime, 58
Chlorine, 14
Chloroform, 64, 148
Chlorosis, 17
Chondrus, 64

220 INDEX.

Chromic acid, 34
Cichorium intybus, 64
Cicuta virosa, 64
Cimicifuga, 64
Cimicifugin, 165
Cinchona, 65
Cinchouiæ sulphas, 65
Cinnabar, 89
Cinnamomum, 65
Cinnamon, 65
    water, 155
Circulatory functions, 191
Citrate of iron, 77
    iron and ammonia, 77
    iron and magnesia, 66
    iron and quinia, 77
    lithia, 98
    potassa, 113
    soda, 66
Citric acid, 64
Cleavers, 81
Clematis erecta, 66
Climate, 8
Climbing staff-tree, 62
Clove-pink, 72
Cloves, 60
Cobweb, 66
Coca, 75
Cocculus indicus, 66
Coccus, 66
Coccyodynia, 19
Cochineal, 66
Cochlearia officinalis, 66
Coco butter, 66, 150
Cocos butyracea, 66
Cod-liver oil, 16, 150
Coffee, 57, 66
Cohosh, 39, 64
Colchici radix et semen, 66
Colchicum, 66
Collinsonia, 67
Collodium, 67, 147
Colocynthis, 67
Coloring principles, 153
Coltsfoot, 134
Columbine, 49
Columbo, 57
Comfrey, 131

Common mallow, 99
Compound resin cerate, 175
    spirit of ether, 169
Comptonia asplenifolia, 67
Concentrated or resinoid extracts, 165
Condition of stomach, 8
Confectio aromatica, 168
    of roses, 168
    opii, 168
    sennæ, 168
Confection, 67, 168
Conium, 67
Constituents of the blood, 204
Contrayerva, 68
Convallaria majalis, 68
    multiflora, 68
Convolvulus panduratus, 68
Copaiba, 68
Copper, 14, 144
Coptis, 69
Corallorhiza odontorhiza, 69
Coral-root, 69
Coriander, 69
Coriandrum, 69
Cornus Florida, 65, 69
Corrosive sublimate, 87
Corydalis formosa, 69
Corylus rostrata, 69
Cotton, 84, 147
Cotula, 69
Cotyledon umbilicus, 69
Cough, 188
Cowhage, 104
Crane's bill, 82
Cream of tartar, 112, 141
Creasote, 14, 70
Creasotum, 70, 147
Creta præparata, 70, 141
Crocus, 70
Croton oil, 150
Crowfoot, 118
Cubebæ, 70
Cuckoo-flower, 60
Cucurbita citrullus, 71
Culver's physic, 96
Cumin seed, 71
Cunilla mariana, 71

INDEX. 221

Cupri acetas, 71
  subacetas, 145
  sulphas, 71, 144
Cuprum ammoniatum, 72, 145
Curcuma, 72
Cuspario, 47
Cyanide of mercury, 88
  potassium, 115
  silver, 50
Cyano-hydric acid, 35
Cyanuret of silver, 50
Cyclamen Europæum, 72
Cydonium, 72
Cynara scolymus, 72
Cynoglossum officinale, 72
Cypripedium, 72
Cytisus laburnum, 72

DAJAKOCH, 72
  Dandelion, 132
Deadly nightshade, 54
Decoctions, 72, 162
Decoctum cetrariæ, 163
  chimaphilæ, 162
  cinchonæ, 162
  cornus Floridæ, 163
  dulcamaræ, 162
  hæmatoxyli, 162
  hordei, 163
  quercus, 162
  sarsap. compositum, 163
  senegæ, 163
  taraxaci, 163
  uvæ ursi, 162
Decubitus of patient, 179
Defect of albumen, 208
  fibrin, 206
  red particles, 205
Defective irritability, 197
  reflex action, 201
  secretion, 202
  sensibility, 198
  tonicity, 197
  voluntary motion, 199
Delphinium, 72
Derangement of sensation, 184
  voluntary motion, 185

Derangements of animal heat, 196
  digestion, 193
  intellect, 186
  perspiration, 195
  reflex system, 185
  respiration, 186
  urinary secretion, 194
Dewberry root, 119
Diabetes, 17
Diachylon plaster, 178
Dianthus caryophyllus, 72
Digestion, 193
Digitalin, 73
Digitalinum, 73
Digitalis, 73
Dill, 46
Dilute acetic acid, 33
  alcohol, 41
  hydriodic acid, 35
  hydrochloric acid, 35
  hydrocyanic acid, 35
  nitric acid, 37
  sulphuric acid, 38
Dioscorea, 73
Dioscorine, 73
Diosma, 56
Diospyros, 73
Distilled vinegar, 32
Dittany, 71
Dock root, 119
Dog's bane, 48
Dog's-tooth violet, 75
Dogwood, 69
Domestic measures, 6
Donovan's solution, 146, 157
Dover's powder, 170
Dracontium, 73
Dragon root, 51
Dried alum, 43
  sulphate of iron, 79
Dulcamara, 74
Dyer's saffron, 60
Dyspnœa, 186

EARTHS, 12, 141
  Effervescing draught, 34
    powders, 170

222　INDEX.

Elaterium, 74
Elder, 121
Elecampane, 92
Elements of disease, 209
Elemi, 74
Elixir of vitriol, 37
Elm, 134
Emetic tartar, 47
Emplastra, 74
Emplastrum ammoniaci, 177
 antimonii, 177
 arnicæ, 177
 assafœtidæ, 177
 belladonnæ, 178
 cum hydrargyrum, 177
 ferri, 178
 galbani compositum, 178
 hydrargyri, 178
 opii, 178
 picis Burgundicæ, 178
 picis cum cantharidæ, 178
 plumbi, 178
 plumbi iodidi, 178
 resinæ, 178
 saponis, 178
Epigæa repens, 74
Epilepsy, 20
Epilobium angustifolium, 74
Equisetum hyemale, 74
Erechthites hieracifolia, 74
Ergota, 74
Erigeron, 75
Erodium cicutarium, 75
Eryngium, 75
Erythronium, 75
Erythroxylon coca, 75
Essences, 168
Essential oil, 104
Ethers, 148
Ethiops mineral, 89
Eucalyptus globosus, 76
Euonymus, 76
Eupatorium, 76
Euphorbia, 76
Excessive irritability, 197
 reflex action, 199
 secretion, 201
 sensibility, 198

Excessive tonicity, 197
 voluntary motion, 199
Excess of albumen, 207
 fibrin, 206
 red particles, 205
Expectoration, 188
Extractive matters, 152
Extracts, 76, 164
Extractum aconiti, 164
 aconiti alcoholicum, 164
 arnica alc., 164
 belladonnæ, 164
 belladonnæ alc., 164
 buchu fluidum. 166
 cannabis purificatum, 164
 cinchonæ, 165
 cinchonæ fluidum, 166
 colchici aceticum, 165
 colocynthidis comp., 165
 conii, 164
 conii alc., 164
 conii fluidum, 166
 digitalis alc., 164
 dulcamara alc., 165
 dulcamaræ fluidum, 166
 ergotæ fluidum, 166
 gentianæ, 165
 gentianæ fluidum, 166
 glycyrrhizæ, 165
 hæmatoxyli, 165
 hellebori alc., 164
 hyoscyami, 164
 hyoscyami alc., 164
 hyoscyami fluidum, 166
 ignatiæ alc., 165
 ipecacuanha fluidum, 166
 jalapæ, 165
 juglandis, 165
 krameriæ, 165
 lupulinæ fluidum, 166
 nucis vomicæ, 165
 opii, 165
 opii liquidum, 166
 podophylli, 165
 pruni Virginianæ fluidum, 166
 quassiæ, 165
 rhei, 165

INDEX. 223

Extractum rhei fluidum, 166
  sarsaparilla comp., 166
  sarsaparillæ, 165
  sarsaparillæ fluidum, 166
  serpentariæ fluidum, 166
  stramonii, 164
  stramonii alc., 164
  sennæ fluidum, 166
  spigeliæ fluidum, 166
  taraxaci, 165
  taraxaci fluidum, 166
  uva ursi fluidum, 166
  valerianæ fluidum, 166
  veratri viridis fluidum, 166
  zingiberis fluidum, 166
Extremities, 184

FACE SYMPTOMS, 180
  False sarsaparilla, 49
    sunflower, 85
    unicorn, 86
Farinaceous medicines, 147
Fel bovinum, 26, 76, 148
Female diseases, 21
Fennel, 80
Fermentum, 76
Ferri bromidum, 144
  chloridum, 77
  citras, 77, 144
  et ammonii citras, 77
  et ammonii sulphas, 77
  et ammonii tartras, 77
  et potassæ tartras, 77, 144
  et quiniæ citras, 77, 144
  ferrocyanidum, 77
  ferrocyanuretum, 144
  iodidi syrup, 167
  iodidum, 77, 144
  lactas, 78, 144
  oxidum hydratum, 78, 144
  phosphas, 78, 144
  pulvis, 143
  pyrophosphas, 78
  ramenta, 79
  redactum, 78
  subcarbonas, 78, 143
  sulphas, 79, 143
  sulphas exsiccata, 79

Ferri sulphuretum, 79
  valerianas, 144
Ferrocyanide of potassium, 115
Ferrum, 79, 143
  ammoniatum, 79, 144
  per hydrogen, 79, 143
Fever-bush, 54
Feverfew, 117
Fever-root, 133
Fevers, 22
Fibrin, 206
Ficus, 80
Figs, 80
Figwort, 123
Filix, 80
Fireweed, 74
Fixed oils and fats, 148
Flaxseed, 97
Fleabane, 75
Florentine orris, 94
Fluid extracts, 166
Fly-trap, 122
Fœniculum, 80
Fowler's solution, 158
Foxglove, 73
Frasera, 80
Fraxinus excelsior, 80
Frostwort, 85
Fucus vesiculosus, 80
Fumaria officinalis, 80
Fumitory, 80
Fusel oil, 41

GALANGAL, 80
  Galbanum, 81
Galega officinalis, 81
Galium, 81
Gallæ, 81
Gallic acid, 35
Galls, 81
Gambogia, 81
Garden artichoke, 72
Garlic, 42
Gases, 14
Gastric juice, 118
Gaultheria, 82
Gelseminum, 82
Genital symptoms, 184

224 INDEX.

Gentiana, 82
    catesbæi, 82
Geranium, 82
German chamomile, 100
Geum, 82
Gillenia, 83
Ginger, 138
Ginseng, 108
Glauber's salts, 127
Glechoma hederacea, 83
Globularia alypum, 83
Glycerate of borax, 174
    carbolic acid, 173
    gallic acid, 174
    tannic acid, 174
Glycerates, 83, 173
Glycerina, 83, 148
Glycyrrhiza, 83
Gnaphalium, 83
Goat's rue, 81
Gold, 13, 83
Golden rod, 127
Goldthread, 69
Gombo, 86
Goose-grass, 81
Gossypium, 84, 147
Goulard's cerate, 177
    extract, 145
Granatum, 84
Grindelia robusta, 84
Ground-ivy, 83
    laurel, 74
    pine, 40
Guaco, 84
Guaiaci resina et lignum, 84
Guaiacum, 84
Guano, 84
Guarana, 108
Gum ammoniac, 44
    arabic, 30
    resins, 151
Gutta-percha, 85

HABITS, 8
   Hæmatoxylon, 35
Hamamelis, 85
Hardhack, 128
Headache, 20

Head symptoms, 180
Heal-all, 67
Heart diseases, 23
Hedeoma, 85
Hedera helix, 85
Hedge garlic, 42
Heleborus fœtidus, 85
Helenium autumnale, 85
Helianthemum, 85
Helleborus, 85
Helonias, 86
Hemidesmi, 86
Hemlock, 67
    pitch, 110
Henbane, 91
Hepatica, 86
Heracleum, 86
Heuchera, 86
Hibiscus abelmoschus, 86
Hickory, 60
Hieracium, 86
Hirudo, 87
Hive syrup, 167
Hoffman's anodyne, 169, 148
Honey, 101
    of borax, 101
Honeysuckle, 98
Hops, 87
Hordeum, 87
Horse-chestnut, 40
Horsemint, 102
Horse-radish, 51
Horsetail, 74
Horse-weed, 67
Hound's tongue, 72
Humulus, 87
Hura, 87
Hydrargyri chloridum corrosivum, 87, 146
    chloridum mite, 88, 146
    cyanidum, 88
    cyanuretum, 147
    iodidum rubrum, 89, 146
    iodidum viride, 88, 147
    oxidum nigrum, 89, 147
    oxidum rubrum, 89, 147
    sulphuretum nigrum, 89, 147
    sulphuretum rubrum, 89, 147

Hydrargyri sulphas flavus, 89, 146
Hydrargyrum ammoniatum, 90, 147
    cum creta, 90, 147
    cum magnesia, 90
Hydrastin, 166
Hydrastis, 90
Hydrate of chloral, 20, 64
Hydrated oxide of iron, 78
    sesquioxide of iron, 78
Hydrochlorate of ammonia, 17, 25, 26, 45
Hydrochloric acid, 35
Hydrocotyle, 91
Hyoscyami, 91
Hyperæmia, 17, 209
Hypericum, 91
Hypophosphite of lime, 19
Hypophosphites, 91
Hyposulphite of soda, 126
Hyssopus, 91

IBERIS AMARA, 91
Iceland moss, 63
Ichthyocolla, 91, 148
Idiosyncrasy, 8
Ignatia bean, 92, 129
Ilex, 92
Impatiens, 92
Imperatoria, 92
Index of diseases, 16
Indian hemp, 48
    physic, 83
    pink root, 128
    sarsaparilla, 86
    tobacco, 98
    turnip, 51
Indigo, 92
Inflammation, 11
Infusions, 92, 155
Infusum angusturæ, 156
    anthemidis, 156
    armoraciæ, 156
    buchu, 156
    capsici, 156
    caryophilli, 156
    cascarilla, 155

Infusum catechu comp., 156
    cinchona comp., 157
    cinchonæ, 155
    calumbæ, 156
    digitalis, 156
    eupatorii, 156
    gentianæ comp., 157
    humuli, 156
    krameriæ, 156
    lini comp., 156
    pruni Virginianæ, 157
    quassiæ, 157
    rhei, 156
    rosæ comp., 156
    sarsaparillæ, 156
    sassafras medul., 157
    senna, 156
    serpentariæ, 156
    spigeliæ, 156
    tabaci, 156
    taraxaci, 156
    ulmi, 156
    valerianæ, 156
    zingiberis, 156
Inhalation, 25, 26, 27
Inorganic products, 139
Insanity, 20
Intellect, 186
Intestinal diseases, 24
Inula, 92
Iodide of ammonium, 46, 92
    antimony, 92
    barium, 92
    cadmium, 56
    calcium, 93
    iron, 77
    lead, 111, 145
    manganese, 99
    mercury, 88
    potassium, 12, 115, 143
    silver, 50, 93
    sodium, 93
    starch, 93
    sulphur, 130, 143
    zinc, 93, 137
Iodine, 12, 93, 143
Iodinium, 93
Iodoform, 93

Iodo-hydrargyrate of potassium, 93
    tannin, 94
Ipecacuanha, 94
Iris, 94
Irish moss, 64
Iron, 13, 79, 199
    filings, 79
Irritability, 197
Isinglass, 91
Isonandra, 85
Ivy, 85

JALAP, 94
    James's powder, 146
Jeffersonia, 95
Jellies, 168
Juglans, 95
Juniperus, 95
    Virginiana, 95

KALMIA LATIFOLIA, 95
    Kamecla, 119
Kidney diseases, 25
Kino, 95
Kooso, 96
Krameria, 96

LABARRAQUE'S DISINFECtant, 140, 158
Laburnum, 72
Lac sulphuris, 134
Lactate of iron, 78
    manganese, 99
Lactic acid, 36
Lactucarium, 96
Ladies' slipper, 72
    mantle, 41
Lappa, 96
Lard, 39
Larkspur, 72
Laryngeal and tracheal diseases, 25
Lavandula, 96
Lavender flowers, 96
Lead, 13, 14, 145
Leech, 87
Lemons, 97

Leontice thalictroides, 96
Leonurus, 96
Leopard's bane, 51
Leptandra virginica, 96
Leptandrin, 166
Liatris spicata, 97
Life-everlasting, 47, 83
    root, 124
Lignin and its derivatives, 147
Ligusticum levisticum, 97
Lily of the valley, 68
Lime, 12, 141
    water, 142, 157
Limones, 97
Linimenta, 97
Liniments, 174
Linimentum aconiti, 174
    ammoniæ, 174
    belladonnæ, 174
    camphoræ, 174
    camphoræ compositum, 174
    cantharidis, 174
    chloroformi, 174
    crotonis, 174
    hydrargyri, 174
    iodi, 174
    opii, 174
    potassii iodidi cum sapone, 174
    saponis, 175
    sinapis comp., 174
    terebinthinæ, 175
Linseed oil, 150
Linum, 97
Lion's foot, 116
Liquids, 172
Liquor ammonia citratis, 157
    ammoniæ acetatis, 141, 157
    arsenici et hydrargyri iodidi, 146, 157
    arsenici hydrochloricus, 157
    barii chloridi, 142, 157
    bismuthi et ammoniæ citratis, 157
    calcii chloridi, 142, 157
    calcis, 142, 157
    calcis saccharatus, 157
    ferri bromidi, 143

INDEX. 227

Liquor ferri citratis, 157
 ferri iodidum, 144
 ferri nitratis, 144, 157
 ferri perchloridi fortior, 158
 ferri subsulphatis, 158
 gutta-perchæ, 158
 hydrargyri nitratis, 158
 iodinii compositus, 142, 158
 lithiæ effervescens, 158
 magnesiæ carbonatis, 158
 magnesiæ citratis, 158
 morphiæ acetatis, 158
 morphiæ hydrochloratis, 158
 morphiæ sulphatis, 158
 plumbi subacetatis, 158
 plumbi subacetatis dilutus, 158
 potassæ, 140, 158
 potassæ arsenitis, 146, 158
 potassæ citratis, 158
 sodæ chlorinatæ, 140, 158
Liquores, 97
Liquorice, 83
Liriodendron, 97
Litharge, 111, 145
Lithiæ carbonas, 97
 citras, 98
Liver diseases, 26
Liverwort, 86
Lobelia, 98
Logwood, 85
Long pepper, 109
Lonicera caprifolium, 98
Loveage, 97
Lugol's solution, 143
Lunar caustic, 50
Lung diseases, 26
Lycopus, 98
Lythrum salicaria, 98

MACE, 104
 Madder, 119
Magendie's sol. of sulphate of morphia, 158
Magnesia, 98, 142
Magnesiæ bicarbonas, 142
 carbonas, 98, 142

Magnesiæ sulphas, 99, 142
Magnolia, 99
Maidenhair, 40
Malambo, 99
Male fern, 80
Malva, 99
Mandragora officinalis, 99
Mandrake, 99, 111
Manganese, 99, 144
Manganesii oxidum, 100
 sulphas, 100, 144
Manna, 100
Maranta, 100, 147
Marigold, 57
Marjoram, 107
Marrubium, 100
Marshmallow, 43
Marsh parsley, 124
 rosemary, 129
Masterwort, 86, 92
Mastiche, 100
Materia medica, 139
Matias, 99
Matico, 100
Matricaria, 100
May apple, 111
Mayweed, 69
Meadow anemone, 46
 saffron, 66
Medicated waters, 49, 155
 wines, 161
Mel, 100
 despumatum, 101
 rosæ, 101
 sodæ boratis, 101
Melia azedarach, 53
Melissa, 101
Menispermum, 101
Mentha piperita, 101
 viridis, 102
Menyanthes, 102
Mercury, 13, 146
 with chalk, 90
 with magnesia, 90
Mesenna, 102
Metallic elements, 143
Mezereon bark, 102
Mezereum, 102

INDEX.

Mikania, 84
Milfoil, 33
Mineral acids, 139
Mistura ammoniaci, 163
   amygdalæ, 163
   assafœtida, 163
   chloroformi, 163
   creasoti, 163
   cretæ, 163
   ferri composita, 164
   glycyrrhizæ composita, 164
   potassæ citratis, 164
Misturæ, 102
Mitchella repens, 102
Modes of death, 213
Momordica balsamina, 102
Monarda, 102
Monesia, 103
Monsel's solution, 158
Moonseed, 101
Morphia, 103
Morphiæ acetas, 103
   murias, 103
   sulphas, 103
Moschus, 103
Motherwort, 96
Mountain laurel, 95
Mucilages, 103
Mucilaginous medicines, 147
Mucilago acaciæ, 30, 103
   tragacanthæ, 103
Mucuna, 104
Mullein, 136
Murias ammoniæ, 45
Muriate of morphia, 103
   quinine, 117
Musk, 103
   artificial, 104
Mustard seed, 125
Myrica cerifera, 104
Myristica, 104
Myrospermum peruiferum, 53
Myrrha, 104

NAPHTHA, 104
   Navel-wort, 69
Neck symptoms, 180
Nectandra, 105

Nervous system, 184
Neutral crystalline principles, 152
   mixture, 164
   organic principles, 152
New Jersey tea, 62
Night-blooming cereus, 56
Nitrate of lead, 111, 145
   potassa, 114
   silver, 21, 25, 50, 145
Nitric acid, 36
Nitro-muriatic acid, 17, 26, 28, 37
Non-metallic elements, 143
Nutmegs, 104
Nux vomica, 105

OAK, 117
   Oatmeal, 53
      gruel, 53
Oil of yellow saunders, 122
Oils, 105, 147
Ointments, 134, 175
Okra, 86
Olea, 105, 149
Oleoresina capsici, 151
   cubebæ, 151
   filicis, 151
   lupulinæ, 151
   piperis, 151
   zingiberis, 151
Oleo-resins, 151
Oleum amygdalæ dulcis, 149
   anisi, 150
   anthemidis, 149
   bergamii, 149
   bubulum, 149
   cajeputi, 149
   camphoræ, 149
   carni, 150
   caryophylli, 150
   chenopodii, 150
   cinnamomi, 150
   copaibæ, 150
   coriandri, 150
   cubeba, 150
   erigantis canadensis, 150
   fœniculi, 150
   gaultheriæ, 150

INDEX. 229

Oleum hedeomæ, 150
  juniperi, 150
  lavandulæ, 150
  limonis, 150·
  lini, 150
  menthæ piperitæ, 150
  menthæ viridis, 150
  monardæ, 150
  morrhuæ, 150
  myristicæ, 150
  olivæ, 150
  origani, 150
  pimentæ, 151
  ricini, 150
  rosæ, 150
  rosmarini, 151
  rutæ, 151
  sabinæ, 151
  sassafras, 151
  succini, 150
  succini rect., 151
  tabaci, 151
  terebinthinæ, 150
  theobromæ, 150
  thymi, 150
  tiglii, 150
  valerianæ, 151
Olive oil, 150
Onion, 105
Opium, 14, 105
  confection, 168
  preparations of, 106, 162
Orange-peel, 53
Orchis, 120
Organic products, 147
Origanum, 107
Orobanche Virginiana, 107
Orris, 94
Ovarian diseases, 22
Oxalate of cerium, 63
  iron, 107
Oxalic acid, 107
Oxalis acetosella, 107
Ox gall, 76, 148
Oxide of manganese, 100
  silver, 51
  zinci, 138,

Oxymel scillæ, 134
Oyster shells, 132

PÆONIA OFFICINALIS, 107
  Pain as symptom, 189
Palm oil, 66
Panax, 108
Papaver, 108
Paralysis, 20
Pareira, 108
Parsley root, 108
Parthenium integrifolium, 108
Partridge berry, 82, 102
Pathology and therapeutics, 197
Paullinia, 108
Pearlash, 113
Pellitory, 116
Pennyroyal, 85
Peony, 107
Pepo, 108
Peppermint, 101
Pepsin, 18, 118
Percussion of chest, 189
Permanganate of potassa, 114
Persimmon, 73
Perspiration, 195
Peruvian balsam, 53
  bark, 65
Perverted secretion, 204
  sensibility, 198
  voluntary motion, 199
Pessaries, medicated, 108
Petroleum, 104
Petroselinum, 108
Pharmaceutical preparations, 155
Phenic acid, 34
Phenol, 34
Phosphas sodæ, 108, 126
Phosphate of ammonia, 45
  iron, 78
  lime, 16, 57, 142
  manganese, 99
  soda, 108, 126
Phosphorus, 14, 109
Physiognomical rugæ, 180
Physiological symptoms, 184
Physostigmatis, 109

Phytolaccæ, 109
Picric acid, 34
Pilewort, 123
Pills, 170
Pilulæ aloes, 171
    aloes et assafœtidæ, 171
    aloes et ferri, 171
    aloes et mastic, 171
    aloes et myrrhæ, 171
    antimonii comp., 171
    assafœtidæ, 171
    cambogiæ comp., 171
    catharticæ compositæ, 171
    copaibæ, 171
    ferri carbonatis, 171
    ferri compositæ, 172
    ferri iodidi, 172
    galbani compositæ, 172
    hydrargyri, 172
    opii, 172
    quiniæ sulphatis, 172
    rhei, 172
    rhei compositæ, 172
    saponis compositæ, 172
    scillæ comp., 172
Pimenta, 109
Pink root, 128
Piper longum, 109
    nigrum, 109
Pipsissewa, 64, 102
Pitch, 110
Pix, 110
    Burgundica, 110
    Canadensis, 110
    liquida, 110
Plantago major, 110
Plantain, 110
Plasters, 74, 177
Platinum, 110
Pleurisy, root, 52
Plumbi acetas, 110, 145
    carbonas, 111, 145
    iodidum, 111, 145
    nitras, 111, 145
    oxidum, 111, 145
Plummer's pill, 171
Podophyllin, 111
Podophyllum peltatum, 111

Poison-oak, 133
Poisonous serpents, 15
Poisons and antidotes, 11
Poke root, 109
Polygala rubella, 111
Polygonum punctatum, 111
Pomegranate, 84
Populus tremuloides, 112
Potash, 140
Potassa, 112, 140
    cum calce, 112, 140
Potassæ acetas, 112, 140
    bicarbonas, 112, 140
    bichromas, 112
    bitartras, 112, 141
    carbonas, 113, 140
    chloras, 113, 140
    citras, 113, 140
    et sodæ tartras, 113
    nitras, 114, 141
    permanganas, 114
    sulphas, 114, 141
    tartras, 114, 141
Potassii bromidum, 115, 143
    cyanidum, 115
    ferrocyanidum, 115
    iodidum, 115, 143
    sulphuretum, 115
Potassio-tartrate of antimony, 47
    iron, 77
Potato fly, 59
Powder of iron, 79
Powders, 169
Prairie dock, 108
Prenanthes, 116
Preparations of earths, 141
    opium, 106, 162
Prepared calamine, 57
    chalk, 70
    honey, 101
Prickly ash, 49, 137
    poppy, 49
Primary elements of disease, 197
Prince's feather, 44
Prinos, 116
Properties of muscular fibre, 197
    nerves, 198
    secretion, 201

INDEX. 231

Proportion of doses, 7
Propylamia, 116
Prostration, 11
Protein and similar principles, 148
Protiodide of mercury, 88
Prunes, 116
Prunum, 116
Prunus Virginiana, 116
Prussian blue, 77
Prussic acid, 35
Pulse, 192
Pulveres aperientes, 170
    effervescentes, 170
Pulvis aloes cum canella, 170
    antimonialis, 146
    aromaticus, 170
    ipecacuanhæ comp., 170
    jalapæ compositus, 170
    rhei compositus, 170
Pumpkin seeds, 108
Purple willow-herb, 98
Purpura, 18
Pyæmia, 17
Pyrethrum, 116
    parthenium, 117
Pyrophosphate of iron, 78

QUASSIA, 117
    Queen's root, 129
Quercus, 117
Quevenne's iron, 79
Quince, 72
Quiniæ murias, 117
    sulphas, 117
    valerianas, 117

RAG WEED, 44
    Raisins, 134
Ranunculus, 118
Rattlesnake-weed, 86
Red-berried trailing whortleberry, 134
Red cedar, 95
    clover, 133
    iodide of mercury, 89
    oxide of mercury, 89
    particles of the blood, 205

Red precipitate, 89
    saunders wood, 122
Reflex action, 199
    or excito-motory system, 185
Rennet, 118
Resina, 118
Resinæ jalapæ, 165
    podophylli, 165
    scammonii, 165
Resins, 151
Respiration, 186
Respiratory changes in blood, 208
Rhamni, 118
Rhatany, 96
Rheum, 118
Rhigolene, 118
Rhubarb, 118
    wine, 161
Rhus glabrum, 118
Rochelle salt, 113
Root of madder, 119
Rosa, 118
Rose honey, 101
Rosemary, 118
Rose petals, 118
Rosmarinus, 118
Rottlera, 119
Rubia, 119
Rubus trivialis villosus, 119
Rue, 119
Rules to proportion the doses, 7
Rumex, 119
Ruta, 119

SABADILLA, 120
    Sabbatia, 120
Sabina, 120
Saccharine medicines, 147
Saccharum, 120, 148
    lactis, 120
Saffron, 70
Sagapenum, 120
Sage, 121
Sago, 120, 147
Salæratus, 140
Sal ammoniac, 45
    prunelle, 141

232 INDEX.

Salep, 120
Salicina, 120
Salix cortex, 121
Salt of tartar, 113
Salvia, 121
Sambucus, 121
Sanguinaria, 121
Santalum, 122
Santonin, 122
Sapo, 122
Saponaria, 122
Sarracenia, 122
Sarsaparilla, 122
Sassafras, 122
Savine cerate, 176
    leaves, 120
Scammonium, 123
Scammony, 123
Scarlet pimpernel, 46
Scillæ, 123
Scoparius, 123
Scouring rush, 74
Scrophularia nodosa, 123
Scullcap, 123
Scurvy, 18
    grass, 66
Scutellaria laterifolia, 123
Secondary elements of disease, 209
Seidlitz powders, 170
Selinum palustre, 124
Senecio aureus, 124
Senega, 124
Sennæ folia, 124
Senna leaves, 124
Sensation, 184, 201
Sensibility, 198
Serpentariæ radix, 124
Serpents, poisonous, 15
Sesamum, 125
Sesquicarbonate of ammonia, 45
Sesquichloride of iron, 77
Sevum, 125
Sex, 7
Silk-weed, 52
Silver, 13, 145
Simaruba, 125
Sinapis, 125

Skunk cabbage, 73
Snake root, 124
Soap, 122
    cerate, 177
Soapwort, 122
Sodæ acetas, 125, 140
    bicarbonas, 126, 140
    boras, 126, 140
    carbonas, 126, 140
    carbonas exsiccata, 126, 140
    et potassæ tartras, 113, 141
    hyposulphis, 126
    phosphas, 126, 140
    sorrel-tree, 46
    sulphas, 127, 140
    sulphis, 127
    valerianas, 127, 140
Sodii chloridum, 127, 140
Solidago, 127
Solomon's seal, 68
Solution of acetate of zinc, 157
    ammonia, 49
Sow-bread, 72
Spearmint, 102
Spermaceti, 63
Spice bush, 54
Spider's web, 66
Spigelia, 128
Spiræa, 128
Spirit of Mindererus, 141, 157
    nitric ether, 158
Spirits or essences, 168
Spiritus ætheris compositus, 148, 169
    ætheris nitrosi, 148, 169
    ammoniæ, 141, 169
    ammoniæ aromaticus, 141, 169
    anisi, 169
    camphoræ, 169
    chloroformi, 168
    cinnamomi, 168
    juniperi compositus, 168
    lavandulæ, 168
    lavandulæ compositus, 169
    limonis, 169
    menthæ piperitæ, 169
    menthæ viridis, 169

INDEX. 233

Spiritus myrciæ, 169
   myristicæ, 169
   pyroxylicus, 41, 147
   tenuior, 41
   vini gallici, 128
Spongia, 128
Spongiæ ustæ, 128
Spunk, 40
Spurge, 76
Spurred rye, 74
Squill, 123
Stanni pulvis, 128
Starch, 46
Star-grass, 42
Starwort, 86
Statice, 129
Stillingia, 129
Stillingin, 165
St. Johnswort, 91
Stomach diseases, 28
Storax, 129
Storksbill, 75
Stramonium, 129
Striped maple, 31
Structural disease, 213
Strychnia, 129
Strychniæ sulphas, 129
Strychnos ignatia, 129
   nux vomica, 129
Styrax, 129
Subcarbonate of bismuth, 55
   iron, 78
Sublimed sulphur, 130
Subnitrate of bismuth, 55
Succinum, 130
Suet, 125
Sugar, 120, 148
   of milk, 120
Sulphate of atropia, 53
   alumina, 44
   bebeerin, 54
   cadmium, 57
   cinchonia, 65
   copper, 71
   iron, 79
   magnesia, 98
   manganese, 100
   morphia, 103

Sulphate of potassa, 114
   soda, 127
   zinc, 138
Sulphite of soda, 127
Sulphocarbolates, 131
Sulphur, 130, 143
   præcipitatum, 130, 143
Sulphuret of iron, 79
   mercury, 89
   potassium, 115
Sulphuric acid, 37
   ether, 40
Sulphuris iodidum, 130, 143
Sulphurous acid, 38
Sumach, 118
Sumbul, 131
Suppositoria acidi tannici, 178
   hydrargyri, 178
   morphiæ, 178
   plumbi compositæ, 178
Suppositories, 178
Sweet birch, 55
   fern, 67
   flag, 51
   spirits of nitre, 169
Symphytum, 131
Symptomatology, 179
   physiological, 182
   topographical, 179
Syrup of blackberry root, 167
Syrups, 167
Syrupus acaciæ, 167
   acidi citrici, 167
   allii, 167
   amygdalæ, 167
   aurantii, 167
   ferri iodidi, 167
   ferri phosphatis comp., 142
   ipecacuanhæ, 167
   krameriæ, 167
   lactucarii, 167
   limonis, 167
   manganesii iodidi, 144
   papaveris, 167
   pruni Virginianæ, 167
   rhei, 167
   rhei aromaticus, 167
   rosæ gallica, 167

234 INDEX.

Syrupus rubi, 167
   sarsaparillæ compositus, 167
   senegæ, 167
   scillæ, 167
   scillæ compositus, 167
   tolutanus, 167
   zingiberis, 167

TABACUM, 131
Tag alder, 42
Tamarindus, 131
Tamarind whey, 132
Tanacetum, 132
Tannic acid, 38
Tannin, 38
Tansy, 132
Tapioca, 132, 147
Tar, 110
Taraxacum, 132
Tartar emetic, 47
Tartaric acid, 39
Tartrate of iron and ammonium, 77
   iron and potassa, 77, 114
Temperament, 8
Terebinthina, 132
Teriodide of formyle, 93
Testæ, 132
Thick-leaved pennywort, 91
Thorn apple, 129
Thoroughwort, 76
Throat symptoms, 182
Thrombosis, 18
Tin, 14, 128
Tinctura aconiti folii, 159
   aconiti radicis, 159
   aloes, 159
   aloes et myrrhæ, 159
   arnicæ, 159
   assafœtidæ, 159
   belladonnæ, 159
   benzoini compositæ, 159
   calumbæ, 159
   cannabis, 159
   cantharidis, 159
   capsici, 159
   cardamomi, 159
   cardamomi compositæ, 159

Tinctura castorei, 159
   catechu, 159
   cinchonæ, 159
   cinchonæ composita, 159
   cinnamomi, 159
   colchici, 159
   conii, 159
   cubebæ, 159
   digitalis, 159
   ferri chloridi, 144, 160
   gallæ, 160
   gentianæ compositæ, 160
   guaiaci, 160
   guaiaci ammoniata, 160
   hellebori, 160
   humuli, 160
   hyoscyami, 160
   iodinii, 160
   iodinii composita, 160
   jalapæ, 160
   kino, 160
   krameriæ, 160
   lobeliæ, 160
   lupulinæ, 160
   myrrhæ, 160
   nucis vomicæ, 160
   opii, 160
   opii acetata, 160, 162
   opii ammoniata, 160
   opii camphorata, 160, 162
   opii deodorata, 160
   quassiæ, 160
   rhei, 160
   rhei et aloes, 160
   rhei et gentianæ, 160
   rhei et senuæ, 160
   sanguinariæ, 160
   scillæ, 160
   serpentariæ, 160
   stramonii, 160
   tolutana, 160
   valerianæ, 160
   valerianæ ammoniata, 160
   veratri viridis, 160
   zingiberis, 161
Tinctures, 158
Toad flax, 48
Tobacco, 131

INDEX. 235

Tolu balsam, 54
Tonicity, 197
Toothache tree, 49
Tormentilla, 133
Touch-me-not, 92
Touchwood, 40
Toxæmia, 17
Toxicodendron, 133
Tragacantha, 133, 148
Trailing arbutus, 74
Trifolium, 133
Trillium, 133
Triosteum, 133
Troches, 167
Trochisci acidi tannici, 167
    bicarbonatis sodæ, 168
    bismuthi, 167
    catechu, 168
    cretæ, 168
    cubebæ, 168
    ferri subcarb., 168
    glycyrrhizæ et opii, 168
    ipecacuanhæ, 168
    magnesiæ, 168
    menthæ piperitæ, 168
    potassæ chloratis, 168
    zingiberis, 168
Tulip-tree bark, 97
Turkey corn, 69
Turmeric, 72
Turner's cerate, 176
Turpentine, 132
Turpeth mineral, 89
Tussilago, 134
Twin-leaf, 95

ULMI CORTEX, 134
  Ulmus, 134
    fulva, 134
Unguenta, 134, 175
Unguentum antimonii, 176
    aquæ rosæ, 175 ;
    belladonnæ, 176
    benzoini, 177
    cadmii iodidi, 177
    cantharidis, 177
    creasoti, 177
    cupri subacetis, 176

Unguentum gallæ, 175
    hydrargyri, 176
    hydrargyri ammon., 176
    hydrargyri iodidi rubi, 176
    hydrargyri nitratis, 177
    iodinii, 176
    iodinii compositum, 176
    mezerei, 177
    plumbi carbonatis, 176
    plumbi iodidi, 177
    potassii iodidi, 176
    simplex, 175
    stramonii, 176
    sulphuris, 176
    sulphuris compositum, 176
    tabaci, 177
    veratri alb., 175
    zinci oxidi, 176, 177
Upas antiar, 72
Upright virgin's-bower, 66
Urinary secretion, 194
Uterine diseases, 21
Uva passa, 134
  ursi, 134

VALERIANA, 135
  Valerian root, 135
Valerianate of ammonia, 45
    bismuth, 55
    iron, 135
    quiniæ, 117
    soda, 127
    zinc, 138
Valerianic acid, 39
Vanilla, 135
Vegetable acids, 153
    alkaloids, 154
Venereal diseases, 28
Veratria, 135
Veratrum album, 135
    viride, 136
Verbascum thapsus, 136
Verbena officinalis, 136
Vervain, 136
Viburnum, 136
Vienna caustic, 112
Vina medicata, 136
Vinegar, 31, 161

Vinegar of bloodroot, 32
  of cantharides, 31, 161
  of lobelia, 32
  of meadow saffron, 32
  of opium, 32, 161
  of squill, 33, 161
  whey, 31
Vinum aloes, 161
  antimonii, 161, 146
  colchici radicis, 161
  colchici sem., 161
  ergotæ, 161
  ipecacuanhæ, 161
  opii, 161
  rhei, 161
  tabaci, 161
  xericum, 136
Viola, 137
Violet, 137
Virginia creeper, 46
Volatile or essential oils, 149
Volume of body, 180
Voluntary motion, 185, 199

WAHOO, 76
  Water avens, 82
    hemlock, 64
    pepper, 111
    plantain, 42
    starwort, 57
Watermelon, 71
Wax, 62
  myrtle, 104
Weights and measures, 5
White hellebore, 135
  horehound, 100
  poppy, 108
  precipitate, 89
  wine, 136
Wild chamomile, 69
  cherry bark, 116

Wild cucumber, 74
  ginger, 52
  indigo, 54
  potato, 68
  senna, 83
  yam, 73
Willow bark, 121
Willow-herb, 74
Wine of aloes, 161
  ergot, 161
  ipecac, 161
  opium, 161
Wintera, 137
Winter's bark, 99, 137
Witch hazel, 85
Wood naphtha, 41, 147
  sorrel, 107
Woody nightshade, 74
Wormseed, 63
Wormwood, 30

XANTHORRHIZA, 137
  Xanthoxylum, 137

YARROW, 33
  Yeast, 76
Yellow jasmine, 82
  root, 90, 137
  wax and white, 162

ZINC, 14, 145
  Zinci acetas, 137, 145
    carbonas præcipitatus, 137, 145
    chloridum, 137, 145
    cyanuretum, 145
    iodidum, 137
    oxidum, 138, 145
    sulphas, 138, 145
    valerianas, 138, 145
Zingiber, 138

THE END.